OUR POLLUTED FOOD

OUR POLLUTED FOOD

A Survey of the Risks

JACK LUCAS

A HALSTED PRESS BOOK

JOHN WILEY & SONS
New York

PUBLISHED IN THE U.S.A.
BY HALSTED PRESS, A DIVISION
OF JOHN WILEY & SONS INC.
NEW YORK

Library of Congress Cataloging in Publication Data

Lucas, Jack.
    Our polluted food.

    "A Halsted Press book."
    Includes index.
      1. Food contamination. 2. Food adulteration and inspection. I. Title.
[DNLM: 1. Environmental health. 2. Food contamination. 3. Food-processing
industry. WA701 L9330]
TX531.L8   1975   614.3'1   75–8700
ISBN 0–470–55285–9

PRINTED IN GREAT BRITAIN

Preface

ENVIRONMENTAL POLLUTION has emerged as a really serious problem only since the end of the Second World War. The scars of the Industrial Revolution and the grosser effects of pollution in the external environment have existed for a much longer time, but it is only within the last thirty years or so, with the increasing magnitude and variety of the problems, that they have penetrated into the public consciousness. The blame for environmental pollution cannot be attributed solely to industrial activities, including power generation; modern agricultural practices, the testing of atomic weapons and the development of the highly efficient car engine must all share some of the blame. The pollution problems are essentially twofold, as they affect firstly the external environment and, in some cases, the pollution is then transferred along a food chain and into man as a result of the consumption of contaminated food.

The worst ravages of external environmental pollution have aroused the most immediate concern and, as a consequence of their often offensive nature, they have already received considerable attention. The problems which arise from the population explosion and the exploitation of the earth's resources, have been the subject of numerous studies and a steady flow of books on the subject. The final consequences of external pollution, with the pollution of the human internal environment, have by comparison received only fragmentary attention to date. The purposes of this book are to review the data on food and human pollution and to attempt a review of the problem in its entirety. It would, of course, be highly satisfactory if the review were able to accomplish a tidy and reassuring answer to the many problems which exist but, unfortunately, this is just not possible on the evidence which is available to date, and there are many problems awaiting further investigation.

The book is not intended to be alarmist, but merely to draw attention to the problems which exist and to promote discussion of them. Many effects which may exist can only be reduced to a very small statistical probability of an effect in man, adding to the small probability of effects which already exist from perfectly natural sources of exposure. This is, of course, no consolation to the victim of a one in a million effect due to exposure to a polluting agent, but even greater concern must relate to the health of future generations for which present humanity does hold a very special responsibility.

This concern has been well expressed by the late American President John Kennedy, in an address to the nation, "the loss of even one human life, or the malformation of even one baby – who may be born long after we are gone – should be of concern to us all. Our children and grandchildren are not merely statistics towards which we can be indifferent" (quoted by E. J. Sternglass, in *Low Level Radiation*, Earth Island Ltd., 1973).

The subject has some technical difficulties but an attempt has been made to present the technical data as simply as possible. The level of presentation has been designed to appeal to students of environmental courses, but it is hoped that the book will also be of interest to members of the Public Health Service and to any members of the general public who have a genuine concern for the future welfare of mankind.

The author is pleased to acknowledge the advice and generous assistance he has received at all times from the Head of his Department at the University, Mr J. C. Collins. The book could not have been completed without his great encouragement. The assistance of other colleagues at work, Dr P. D. Day of the Chemistry Department and Dr P. J. W. Saunders of the Pollution Research Unit, is also gratefully acknowledged. Any deficiencies which the present book may suffer from are, of course, entirely attributable to the author. The author is also extremely grateful to Mrs Ann Bieri, Mrs Elaine Watmough and Mrs Collinson for their understanding and help in the preparation of the manuscript.

The author is grateful to Mr Robert Waller (consultant editor of Charles Knight's ecology books), for his invitation to undertake the task of preparing this book.

The data which have been used to assess the dietary intakes of the various pollutants have been obtained mainly from official U.K. reports and review articles with the emphasis on the secondary sources of information at the expense of the primary sources to be found in original papers. This approach is illustrated by the use of the International Commission on Radiological Protection, *Publication 2*, 1959, for data on the distribution of certain elements in human tissues, the data in this publication having been obtained from the comprehensive studies of Isabel H. Tipton and her colleagues, whose recent results have been published in the journal *Health Physics* 1963 **9** pp. 105–33. The author is conscious of the debt he owes to many such workers and makes grateful acknowledgement to them all, especially to the undermentioned, for permission to quote at length or to make extensive use of published data:

The Controller of Her Majesty's Stationery Office, London, for data from several series of official reports; the Agricultural Research Council, Letcombe Laboratories, Wantage, Berks; Ministry of Agriculture, Fisheries and Food, Fisheries Radiobiological Laboratory, Lowestoft; the Association of Public Analysts with Local Authority Organisations, London (Joint Surveys); the Editorial Services of the World Health Organisation, Geneva, for several series of official reports; the California State Department of Health, Berkeley, California; the US Government Printing Office (report of the Secretary's Commission on pesticides and their relationship to environmental health); the Clarendon Press, Oxford, for the poem "Cologne" by S. T. Coleridge; Academic Press (London) Ltd.; W. H. Freeman & Co., San Francisco, California, publishers of *Scientific American*; Microforms International Marketing Corporation, Flinford, N.Y., and H. A. Schroeder (for data from several series of monumental surveys of trace elements in diet published in the *Journal of Chronic Diseases*); New Scientists Publications, London; Macmillan Journals Ltd., London, and H. William Smith for articles in *Nature*; and Henry Doubleday Research Association with the Michaelis Nutritional Research Laboratories for data on lead.

Finally, the author wishes to thank the artists Paul Bailey and Ann Savage for their work in preparing the series of figures in the text and for their skill in interpreting the author's requirements.

Contents

List of Figures

There once was a physicist: Lucas
Who warned us of ruin – and shook us
With threat of diffusion
Of pest and pollution
And all the foul brews they could cook us
Thomas Pitfield, 1971

I

Basic Principles

INTRODUCTION

AT THE moment, prophets of doom forecasting the end of civilisation within, perhaps, a decade, or no later than the end of the century are many and vociferous. The pollution lobby has witnessed a spectacular growth in this last decade and has even become a fashionable political issue in most of the developed nations. This political concern reflects the growing awareness and mounting anxiety of the general public and the increasing attention which is being given to pollution problems in the news media. A random example of this attention is illustrated by headlines in the pages of the *Daily Telegraph* of 29 January 1971.

1. "Lead poisoning danger to child scavengers"
2. "Tests to beat agricultural pollution"
3. "Orchid grower says jet fumes kill his stock"
4. "Fire risk drums put on tip"
5. "Radioactive waste in Welsh bread"
6. "Lead scare – plant suspends work"
7. "Dismay over toxic waste scandal"

Political concern is also indicated by the presence of strong government delegations at the United Nations Conference on the Human Environment in Stockholm, 1972.

The author does not count himself among the prophets of doom, having every confidence that man is a sufficiently rational being to acquire the necessary knowledge, both to understand his present predicament and the will to overcome the problems which undoubtedly do confront him at the present moment. It is, however, clear that mankind will need to demonstrate a collective wisdom which has not yet been too much in evidence if acceptable solutions to the many problems are to be found. The backing of governments at a national level and their co-operation at the international level, and supported by informed and responsible public opinion will also be essential to achieve the necessary progress. There is also a real danger that problems will increase in severity unless some effective counter-measures are undertaken fairly quickly, and there is now refreshing evidence that some of these measures are being taken.

The problems of the human environment to which attention is increasingly drawn are mainly the overt and, all too often, obvious consequences of the over-exploitation of the earth's resources, some of which are already recognised to have a very limited lifetime. These external effects on the environment include the worst features of urbanisation, with industrial and mining operations causing gross problems of pollution by their products and wastes such as oil spillages in the oceans, chemical waste discharges into rivers and unsightly slag heaps. The problems of increasing noise levels, some deterioration of once fertile lands into dust bowls, atmospheric pollution from car exhaust emissions and the problems of domestic refuse, human and animal sewage are all additional important forms of pollution. It is generally agreed that the impact of these various sources of pollution has most often been harmful, leading to a deterioration in the quality of the human environment.

The subject of this book is, however, concerned with another aspect of the pollution problem which is generally much more insidious and much less obvious. It concerns the problems of internal human pollution which are the end result of the pollution of the external environment.

The nature of this problem has been admirably expressed by the British Government in its submission to the UN Conference on the Human Environment in Stockholm, 1972.[1]

Pre-eminent among the scientific problems that need to be confronted are:

1. The identification of substances which could be harmful if emitted into the environment or concentrated in animal and plant tissues long before they reach levels at which damage begins. This calls for a better early warning system than exists at the moment and a basis of scientific knowledge which allows the behaviour of substances to be predicted from their chemical structure.

2. Better knowledge of the effects of pollutants, especially those resulting from long-term exposure to potentially harmful substances at concentrations which are not dangerous in the short term.

3. Better understanding of the pathways of pollutants in the environment and the rate at which they are removed or destroyed.

4. The measurement of the ecological effects of minor changes in the balance of the environment, where these are sustained over long periods. Those which might affect world climate or the cycles by which oxygen, ozone or carbon dioxide in the air, or nitrogen in the soil are maintained, are of special importance.

All these problems demand, if rational action is to be taken, much more knowledge of the relationship between the concentration of a pollutant in the environment and its effects. Such knowledge is vital if risks are to be assessed. But the whole question of what constitutes an "acceptable risk" bristles with the need for assessment.

The theme of this book therefore concerns the external environment only as the source of the many pollutants moving through a complex series of food chains into the food consumed and thence into man's own internal environment. In this sense the pollution of man himself is the culmination of the despoliation of the outer environment and it is within man himself that the problems generally reach their maximum complexity. This complexity is associated not only with the number of potentially harmful agents, but also with the delicate balance (or homeo-stasis) of the whole range of metabolic processes within the body and the many possibilities which arise for interference in these processes by polluting agents. Many of these agents are total strangers to man's internal environment as they are not present naturally in the external environment and others are increasing in concentration to levels beyond those which are found to occur naturally. In this situation, therefore, there are certainly no grounds for complacency and a great deal of cause for genuine concern for the whole future health of mankind. Above all it is necessary to increase human understanding and knowledge of the problems of multiple exposures and to strive for a measure of perspective in assessing the hazards to health which may arise.

THE HISTORICAL VIEW

THE PRESENT human situation should be viewed in an historical perspective. Mankind today, perhaps not for the first time, finds himself in an age of great changes, many of which are having a profound effect on the whole human environment. Some of these changes may be identified with the rapid growth in the population, and with revolutionary changes in industry, food technology, and in agriculture all to meet the increasing human demands. Some of these changes are of very recent origin and do carry a new threat to the environment. On the other hand certain other changes over a somewhat longer time span do offer a measure of encouragement for the future.

The English poet and philosopher, Samuel Taylor Coleridge, was a fairly frequent visitor to Germany in the early 19th century, being a great admirer of the school of German philosophers. After one such visit he was moved to compose the lines "Cologne" in 1828.

In Köhln, a town of monks and bones,
And pavements fang'd with murderous stones
And rags, and hags, and hideous wenches;

I counted two and seventy stenches,
All well defined, and several stinks!
Ye Nymphs that reign o'er sewers and sinks,
The river Rhine, it is well known,
Doth wash your city of Cologne;
But tell me, Nymphs, what power divine
Shall henceforth wash the river Rhine?

Somewhat later it was not uncommon for sessions of the Houses of Parliament to be abandoned owing to the stench from Old Father Thames; urchins in the city of Bradford could throw lighted matches on to the surface of the canal, lighting a flame which would travel some distance along the water, enveloping any barges and persons in its track. All these happenings were due to a single cause – untreated and rotting human sewage, disposed of directly into the most convenient surface water. Apart from the despoliation of the environment and the obvious offensive nature of such activities, the risk of polluting supplies of drinking water and more indirectly food by pathogenic germs was very real. This form of pollution was undoubtedly a major cause of several plagues affecting Europe in earlier times. Greater knowledge of the dangers to health and the demand for improved hygiene were largely responsible in the latter part of the 19th century for the development of measures to treat sewage before disposal. Although it is not possible to assert that this problem is completely solved, as direct disposal into the seas around the coasts of the country and dumping into the deeper reaches still take place, it is possible to claim that considerable progress has been made in dealing with this form of pollution. Samples of fish have returned to the lower reaches of the Thames for the first time in over 100 years. The situation today, however, is that environmental pollution has become more varied and complex, with very many more chemicals involved. The incident which occurred in the river Rhine in the summer of 1969 provides an apt illustration. A drum of the powerful chemical insecticide Endosulfan fell into the river and despite the exceptionally low levels of concentration of the insecticide in the river water due to the great flow and vast dilution in the river, the chemical was reconcentrated in fish with disastrous consequences to the fish populations downstream. This incident in the river Rhine serves to emphasise the changing nature of the whole environmental pollution problem.

THE POPULATION EXPLOSION

ALTHOUGH VARIOUS forms of pollution have been an inevitable consequence of all human activities since man began to dominate the

earth, the magnitude and complexity of the problems today is undoubtedly due to the ever more rapid growth of the human population. This
growth is illustrated by the figures in Table 1.1 and can be appreciated

Table 1.1
Population Growth

Year	Population	Doubling time
1830	1 billion (i.e. 1,000 million)	100 years
1930	2 billion	45 years
1975 estimate	4 billion	

best from the rapidly decreasing interval of years during which the
world population is doubling. Associated with this explosive growth of
the world population there is also a change in the distribution of the
population. This change is illustrated by figures for the UK which
show that in 1801 only about a quarter of the total population of 10
million lived in the towns and cities whereas 100 years later the figures
had been completely reversed with only about a quarter of the total
population of 33 million then living in the countryside. This trend
continues at a decreasing rate in the Western nations but is probably
only just starting in the developing countries.

Many problems derive from the population explosion and urgent
attention is being given to ways and means of slowing down the growth.
The rapid increase in the present century is related to improvements in
medical practice and public hygiene reducing infant mortality and
raising the average expectancy of life to a greater age. The methods of
slowing down the growth rate based on birth control raise complex
issues and it is unlikely that any rapid improvement in the situation
can be expected for some years ahead. An "International treaty for the
non-proliferation of children" must remain just a hope for the future,
and it is clear that attention must not be diverted from the problems of
feeding the extra population and controlling global pollution.

The population explosion is a decisive factor in aggravating the
pollution problem; it does this by applying pressures on agriculture
and modern industry to supply the needs of the increasing population.

1. The pressure on agriculture is also a pressure on the land available
for cultivation and is generated by the need not only to feed the increasing population but to raise the nutritional standards of about two-thirds
of the population living in the developing countries. Recent surveys by
FAO reveal the full extent of under-nutrition and malnutrition in these
countries. There is, therefore, a tremendous pressure on any land,
which is not being swallowed up by urban sprawl, industrial developments and roads, and on all agricultural resources to improve productivity. This has led inexorably to the increasing use of chemicals, many

of which have created pollution problems. The tremendous increase in the applications of chemicals to improve soil fertility and to afford crop protection, coupled with changing patterns in crop growing, have accelerated external environmental changes and have also led to human internal pollution by these same chemicals, many of which are absent from the natural environment. The accelerating change to intensive stock rearing or "factory farming" has produced another set of problems (see Chapter 3).

2. The pressure on industry is to meet the demands of the people for energy and for other material goods of civilisation such as the car and manufactured products for the modern home. All the evidence appears to indicate that the demand for energy in the developed nations is growing at an even faster rate than the growth in population. In the industrial field the major polluters of the environment are the power industries, such as electricity, gas, coal and oil, and the major chemical industries such as petroleum products and plastics, iron and steel, non-ferrous metals, cement, paper and fertilisers. The waste products from these and other industrial concerns are many and varied. In some cases there have been improvements in controlling the worst features of industrial waste discharges into the environment but in others the effects have been mounting since the industrial revolution over one hundred years ago. They have been with us longer than the comparatively recent agricultural revolution involving the application of chemical products. The growth of industrial pollution has continued despite the introduction of the Alkali Acts before the present century and demonstrates only too well that even legal constraints are inadequate to control serious pollution if the public will is lacking. Far greater success has attended the introduction of the much more recent Clean Air Acts and the control over domestic fuel consumers by the designation of smokeless zones. Similar constraints may be expected in the future to control the worst features of toxic discharges by the modern motor car.

National defence based on a nuclear strategy and programmes for the exploration of near and far space are also potential sources of environmental pollution and are also included in the industrial section. The origins of the various sources of pollution and the more probable food carriers for man are summarised in Table 1.2.

POLLUTION (Definitions)

POLLUTION has been defined as the deliberate or accidental contamination of the environment with man's waste.[2] The total number of pollutants in waste which are now entering the environment is very large, the examples given in Table 1.2 representing only a very small sample of the total. It cannot, however, be assumed that all waste

Table 1.2
Food pollution sources

Source	Type	Examples	Animal produce (fish)	Fats	Vegs	Fruits	Cereals
Agriculture	Pesticide residues	Herbicides; 2, 4-D, 2, 4, 5-T	X		?		?
		Fungicides: Cu, Hg			X	X	X
		Insecticides: DDT	X	*	X	X	
	Fertilisers	Nitrate	X	X	X	X	
	Intensive animal husbandry	Antibiotics: Penicillin	X				
		Hormones: Oestrogens	X	X			
		Micro-organisms	*				
Industrial	Nuclear power / Atomic weapons	Radioactive fall-out Sr90, Cs137	*		X	X	*
	Chemical wastes	Cd, methyl mercury, PCBs	*				
Domestic	Car exhausts	Pb			X	X	X

* Indicates major source

7

products are necessarily harmful either to the environment or to man. Sir Kenneth Mellanby has suggested an alternative definition that "man-made pollution occurs when some activity causes damage". He considers it important that the definition should specify that "damage must be done".[3] This book is concerned with the human effects of pollution and a pollutant is therefore defined as any waste matter (or energy) which causes damage to the environment and may lead to an adverse effect on man. This also creates problems as most of the evidence for harmful effects is obtained at much higher pollutant concentrations than those at present prevailing in man or the environment. Inferences of damage at low concentrations are difficult and frequently tendentious. Even if damage cannot be directly demonstrated, it must therefore be a reasonable inference that harmful effects are probable. Sir Kenneth Mellanby has also pointed out that many instances of pollution are likely to arise in future simply due to the skill of the chemist in detecting and measuring small quantities of pollutants at levels which have so far escaped detection and are highly unlikely to lead to any harmful consequences.

The complexity and seriousness of pollution problems today is very largely due to the insidious nature of the variety and number of potentially harmful agents. It must also be recognised that many of these agents may interact with man and his environment in two distinct ways, externally and internally.

EXTERNAL EFFECTS OF POLLUTION

THE EXTERNAL mode of damage is harm which is done directly to the environment external to man. The harm done to the environment may affect the quality of man's surroundings by a loss of amenity value. In this case the damaging effect occurs primarily in the environment and has only secondary consequences for man himself. The incident which occurred in the Rhine provides a good illustration of the external effect. The destruction of the fish life in the river represented an undesirable state of affairs with an obvious loss in the quality of the river environment, in addition to a loss of amenity in the form of the recreational facilities normally enjoyed by anglers. This incident suggests a further possibility that certain agents may exert external effects through a reduction in the productivity of the environments on which food supplies depend. This is not only evident when severe but local chemical pollution has occurred, but may also be happening to some degree on a much larger scale owing to low but persistent concentrations of certain agents in the environment. One example is sulphur dioxide released into the atmosphere through the combustion of oil in oil-fired electric power stations. It is directly harmful to man at quite low concentrations in the air, and has been one factor responsible in the past for much individual suffering and deaths during smog conditions. For this reason, it has been customary in more recent oil-fired stations to discharge the products of combustion into very high chimney stacks which results in a much greater dilution of sulphur dioxide in the atmosphere and its conversion into sulphuric acid before reaching ground level. The concentrations at ground level, especially in the immediate environment of the power station, are substantially reduced and any harmful effects are mitigated. The total quantity of sulphur dioxide which is emitted by the combustion of oil is, however, on the increase and the effect of dispersal by high chimney stacks may be transferring the ultimate problem elsewhere with possible long-term effects on crops and afforestation. Similar problems may exist for releases of grit and fluoride and in all these cases the reduction in crop yields may be related simply to the amount of deposition and damage to leaf surfaces. Mercury and other chemicals in sea waters may also inhibit photosynthesis by plankton which, in turn, limits the available food supplies of fish, and hence reduces the harvest of the seas.

Yet another type of external effect for man arises with artificial radioactive isotopes when these are deposited in the human environment. The radiations emitted by these substances are often quite penetrating and can contribute a small additional dose of radiation to man. The external effect for radioactive materials is, however, generally far less serious than any internal effects, which are discussed in Chapter 4.

There are also many examples of gross environmental pollution which produce serious external effects for man. They include such problems as oil on the seas and on the beaches, the chemical pollution of the rivers, the scars of mining and manufacturing industry, and urban squalor in general. These gross effects are immediately obvious and, quite rightly, attract a great deal of attention. They have been the subject of many studies and reports but are not directly the subject of the present book. Agents from these and other sources which are polluting the environment can give rise to the second major problem – the pollution of food and man's internal environment.

INTERNAL EFFECTS OF POLLUTION

THE EXTERNAL effects of pollution have only an indirect and perhaps incalculable effect on human health, mainly by reducing the capacity of the environment to satisfy the essential needs for recreation and enjoyment and the productivity of the normal sources of food. The situation may be very different in the case of internal pollution, which involves a transfer of the pollutants from the environment to food and drinking water and thence to man. Apart from the food and water consumed, the polluting agents may also enter the body by inhalation of contaminated air or by direct contact of the skin with contaminated materials. Inhalation and skin contact are much more of an occupational hazard in certain chemical and manufacturing industries and are unlikely to be serious sources of exposure to the general public. Lead from engine exhausts and volatile fission products (see Chapter 4) such as iodine-131 and krypton-85 may also be exceptions in the case of inhalation, but in the case of I-131 contaminated milk and, to a lesser extent, drinking water will still be the major concern. Drinking water in the UK is maintained at exceptionally high levels of quality and barring major disasters it is unlikely that it will assume the importance of contaminated food. The latter is thus the major source of internal human pollution and as such is the major concern of the subsequent chapters of this book.

The contamination of food is the final stage of a sequence of events starting with the discharge of the agent and its subsequent movement through various levels of the environment into the food. Whenever a particular route or critical pathway for the contaminant terminates in a food consumed by man this is referred to as a food chain. Many examples of food chain will be discussed in the subsequent pages, and they will be found to vary considerably in their complexity. Whenever a food chain exists, there is either the possibility that the concentration of the contaminant will increase as it moves through the chain, or, alternatively, it might be discriminated against with obvious benefit to man. The processing of the food into an edible product may produce further changes in the levels of contamination, but in all cases a contaminated product is the end of the chain and hence becomes one of the sources of human contamination. The contaminating agent may then end up in the tissues of the body and may have consequences for the internal body environment.

There is yet another form of contamination which involves a variety of micro-organisms which may be carried in food and drinking water as the primary source of human infection. These micro-organisms include several strains of pathogenic bacteria such as salmonellae and viruses which may be responsible for the spread of infection through their presence in food and drinking water. The greatest danger to drinking water supplies is today generally recognised to be the possibility of contamination by faecal bacteria from sewage. Food as a factor in the spread of certain diseases, such as enteric fever, dysentery, cholera, diphtheria, poliomyelitis and scarlet fever is, however, also well established. Food and water, as a result of careless handling, may also carry pathogenic germs of faecal origin; these may be responsible for severe infections such as typhoid and for the various forms of food poisoning.

In these cases a breakdown in the standards of personal hygiene in the handling of food is often the basic cause of the food contamination and its transfer to other individuals. The bacteria which are responsible for food poisoning in man are also shared with his domestic and farm animals so that the later stages of food chains may also be involved in the transfer of the pathogenic organisms. These problems are given special attention in Chapter 3.

ACUTE AND CHRONIC TOXICITY

THE MICRO-ORGANISMS responsible for food poisoning are very widely distributed in man and his domestic animals. The symptoms of food poisoning are only likely to appear, however, if the supply of drinking water or food carries sufficient numbers of the bacteria. It appears that there must at least be an invasion of the body by a certain minimum number of organisms before their multiplication in the body leads to sufficient numbers for toxic symptoms to develop. A similar situation exists in all probability, with respect to, the quantities of a great majority of the chemical pollutants which find their way into food. Even when it is well established that a substance is a chemical poison, there still exists a minimum dosage level above which harmful effects will appear rapidly. It does not always follow that harmful symptoms or that any form of damage will result when the exposure is limited to very low concentrations which may occur over extended periods of time. The upper limits for the concentrations in drinking water of some well-

Table 1.3

Recommended upper limits of concentrations for selected substance in drinking water
(based on WHO European Standards for Drinking water[4])

Cationic		Anionic and miscellaneous	
Ion	μg/ml	Substance	μg/ml
Cd^{2+}	0·01	Se	0·01
Pb^{2+}	0·1	As	0·05
Cu^{2+}	3·0[p]	Cr	0·05
Zn^{2+}	5·0[p]	CN^-	0·05
Fe	0·1[p]	F^-	1·3
Mn	0·05[p]	NO_3^-	100
		Anionic Detergents	0·2
		PAH*	0·0002
		H_2S	0·05[p]

[p] Levels in excess of the quoted valued may not be toxic to man but may affect the potability of the water
* Polyaromatic (polycyclic) hydrocarbons

known toxic substances are listed in Table 1.3 as examples of "safe" concentrations.

It is important, therefore, to maintain a distinction between what are described as the acute and chronic effects of exposure. The former involves the intake of a relatively massive dose of an agent in a short period of time leading fairly rapidly to well-known symptoms of poisoning or even fairly rapid death. The chronic form of exposure on the other hand involves repeated exposures to very small doses of a toxic substance which are received over a long period of time. In these cases the damage which may occur on a minute scale is either repaired by the normal mechanisms of the organism and the symptoms which are typical of the acute exposure do not occur or any effects may remain latent only to be expressed at a much later date. The values quoted in Table 1.3 are the recommended levels for chronic exposure at which it is confidently predicted that no ill effects will occur during the normal life-span of the individual. This concept of safe levels of exposure may carry with it the principle of a threshold level below which no harm will result, but above which the risk of harmful effects increases with the amount of the dose. This principle may be valid for many of the agents dealt with in the following pages but it cannot always be regarded as true and especially for those agents which accumulate within the organism or which cause cumulative and irreparable damage. The damage to cell nuclei by ionising radiations is an example of the latter. Agents of this nature must, therefore, be regarded as the more hazardous types and their effects at low doses require the most careful consideration and evaluation. All the recommended concentrations in Table 1.3 incorporate safety factors so that harmful effects are unlikely to occur even at levels which may be several times greater than those which are listed. Very much higher concentrations would be necessary to produce rapid, acute symptoms of damage.

The chronic effects of exposure to a variety of polluting agents are the real concern of this book. They are generally very difficult to quantify at the very low concentrations prevailing in food and in man. Their effects may be characterised by a long period of latency and their symptoms may not even be expressed in the life-time of an individual. They are of even greater concern when they affect the germ cells of the body, when their harmful effects may not be expressed until future generations. The nature of the chronic effects both somatic and genetic, is analysed more fully later in this chapter.

LIFE CYCLES AND FOOD CHAINS

NATURE and the whole universe are in a state of constant activity This dynamic character is illustrated by the cyclical nature of all natura processes from the extremely rapid pulsations of infinitely small atoms (used in the caesium clock) to the huge time-scale of the very long-term rotation of the galaxies. There are also many cycles within nature which support and indeed make human life possible. Man is the apex of a pyramid of life forms, and is supported by and also directly involved in a broad base of cyclical and interlocking processes beneath him. The disturbance of the cycles, in many cases unwittingly, may be having profound effects on the quality of human life in very many ways.

The freshwater cycle in Fig. 1.1 is an example which is of great importance in ensuring the quality of surface waters both for use as

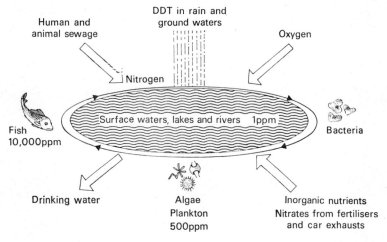

Fig. 1.1 The basic cycle within the freshwater environment. Concentrations refer to DDT at the various trophic levels within the eco-system

drinking water and as a source of the small amount of freshwater fish in the human diet. The maintenance of this quality is also vital for the preservation of the natural environment for human enjoyment and recreation. Waste matter excreted by fish and diluted by the water is converted by micro-organisms with the assistance of dissolved oxygen into simple inorganic nutrients such as nitrates. These in turn feed the algae and other plant forms which are then consumed by the fish which also require a supply of oxygen. The net result of the cycle is normally a balanced system in which all the organisms establish themselves at levels at which they participate most effectively in the life cycle.

Although such cycles are self governing and maintain a dynamic but steady state condition of indefinite duration a variety of human activities are now interfering with the steady state condition as represented in the figure.

Animal wastes from the intensive stock rearing units and human wastes reaching surface waters increase the bacterial activity and the consumption of oxygen. Once the oxygen is consumed, the cycle is interrupted and the bacteria and the fish die. The application of artificial fertilisers to soil in quantities greater than the needs of the growing plants leads to run off from the soil and entry into the cycle; eutrophication with massive overgrowths of algae and their eventual decay may then be the consequence, again leading to excessive oxygen demand and interruption of the cycle. The application of pesticide chemicals in agriculture also leads to their entry into the cycle with concentration stages occurring from water to plants to fish and beyond the cycle to birds and animals feeding on the fish. This action may be damaging to plant, fish and bird life dependent on the fresh water cycle.

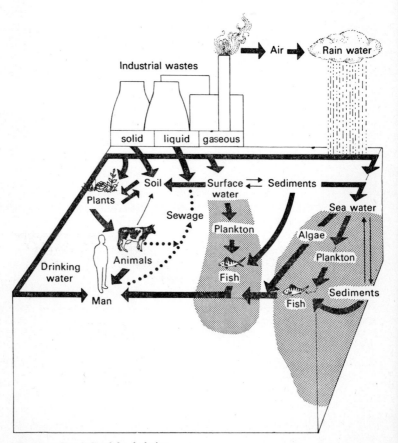

Fig. 1.2 Generalised food chains

The fresh water ecosystem is just one example of many cycles in nature illustrating how pollution of various forms can have external environmental consequences. These external pathways of the polluting agents in the environment are also of considerable importance in determining those foods in which an agent is preferentially concentrated and which then become particularly hazardous for human consumption. A food chain or critical pathway is the sequence of stages in which a hazardous substance is transferred through various levels of the environment terminating in the food product. It will be evident that a great variety of possible pathways exists and that the identity of the critical food chain is clearly of considerable importance for devising effective measures of control. Generalised food chains for the three principal types of environment involved in food production – the agricultural, fresh water and sea-water environments – are represented in Fig. I.2.

Those pathways which include one or more stages of concentration of a pollutant will be of crucial importance. These concentration stages may exist at the boundary between two trophic levels of the chain, e.g. soil to plant, or they may exist internally within an organism, as in the

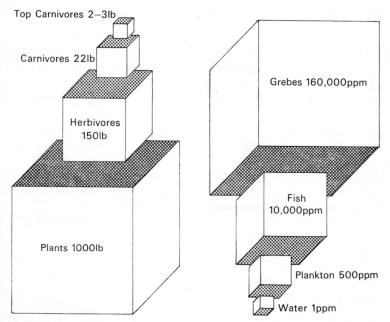

Fig. 1.3 Food pyramid and inverted pyramid for DDT concentrations: (*left*) protein productivity in a pond based on data of N. W. Pirie;[5] (*right*) DDT in freshwater environment

passage from bloodstream to milk or bloodstream to flesh of an animal. The importance of concentration factors is further emphasised when it is remembered that there is a significant reduction in the yield of edible food in progressing from lower to higher levels of a food chain. The food yield can be expressed in terms of an essential nutrient and is illustrated in Fig. 1.3 in the form of the familiar food pyramid for protein production at the successive trophic levels in a fresh water pond. When pollutant concentration factors operate in a similar environment, the concentration with respect to the edible food will increase progressively from the lower to the higher levels, resulting in an inverted pyramid; such a pyramid for DDT concentrations in a fresh water environment is also shown in Fig. 1.3. The results can also be expressed arbitrarily as a ratio of DDT concentration over protein production, Table 1.4. These values cannot be interpreted in any absolute sense

Table 1.4

Concentration of DDT in relation
to protein yields in a possible food chain

Trophic level	Protein Yield[5] lbs	DDT Concentrations ppm	Ratio ppm DDT/lb Protein
Original Plants	1000	500	0·5
Carnivores	22	10,000	450
Top Carnivores	3	160,000	57,000

but merely serve to emphasise the potential seriousness of the situation in food whenever concentration stages can occur. The reverse situation operates when discrimination against the contaminant occurs in moving from a lower to a higher level and this is then to the obvious benefit of the higher organisms.

FOOD CONSUMPTION

THE DAILY human intake of any polluting agent will depend on its concentrations in the various foods consumed and on the amounts of these foods in the daily dietary. The estimates of the daily intakes of the various substances which are reported in the later chapters are based on British statistics for average food consumptions, published in the Annual Reports of the National Food Survey Committee.[6] The data in the 1971 Report covering the years 1969–70 have been arranged in the same Food Groups used in the Total Diet Studies for certain pesticide residues as reviewed in Chapter 2, and are summarised in Table 1.5.

Table 1.5

Average daily intakes of food, UK, 1969–70 [6]

Food group	kg/day	% of Total diet	% of Total diet USA 1965*
1 Cereals and cereal produce	0·25	16	15
2 Fruits, preserves (incl. sugar)	0·20	13 ⎫	
3 Root vegetables	0·26	17 ⎬	35
4 Green vegetables	0·10	7 ⎭	
5 Meats, incl. fish, eggs	0·21	13·5	20
6 Fats	0·05	3	3
7 Milk, cheese	0·45	29	27
8 Miscellaneous†	0·05	1·5	–
9 Totals	1·55	100	100

*Based on estimates of the US Department of Agriculture, Consumer and Food Economics Research Division 1965
†Includes tea, coffee, etc in beverages, ice cream, pickles, sauces
N.B. The British intakes of food have not varied significantly from about 1964 to 1970; the % intakes have been rounded off to provide average estimates representative of the period. There are variations in the average intakes related to different income groups and to regional variations in dietary habits.

The % contribution of each group to the total diet, column 3 has been used in the subsequent chapters to calculate the daily intakes of the various contaminants. The data for food consumption in Chapter 2 for calculating the dietary intakes of pesticide residues are however taken direct from the official report. They differ slightly from the values of column 3 but the differences are unimportant in calculating the intakes.

It has to be emphasised that the data presented in Table 1.5 represent average national values of food consumption, and that there will be considerable individual variations depending on personal circumstances. There will also be variations arising from differences in the patterns of food consumption in the regions of Britain. It is to be expected therefore that actual personal consumptions of various pollutants will show considerable variation from the average values which are calculated from average figures of food consumption. These differences depending on regional and on personal factors will mean that all estimates of intake based on the national food consumptions will provide only a general guide to the levels of intake in the whole community, and some allowances must be made for regional and personal variations. Average values of the concentrations of pollutants in food have also to be used, introducing further uncertainties into the calculations.

PATTERNS OF ABSORPTION AND DISTRIBUTION IN THE BODY

THE ABSORPTION and distribution of any substance entering the body in food and its ability to cause damage will depend essentially on three factors.

1. The metabolic pathway of the substance within the body, the chemical transformations which it may undergo at various stages, and the rate at which it is eliminated.
2. The organ or tissues of the body in which the original substance or an active metabolite is preferentially concentrated.
3. The relative sensitivity to damage of any tissues exposed to significant levels of the substance.

The essential metabolic pathways for substances ingested in food or water are represented diagrammatically in Fig. 1.4. Absorption of any substance in its passage through the gastro-intestinal (g.i.) tract depends on a number of intrinsic physical qualities such as its solubility in water or in fats, and the magnitude of any electrical charge on the molecules or ions, as the smallest units of the substance. Absorption is also influenced by the quantity and the quality of the food consumed and to some extent by populations of micro-organisms inhabiting the lower parts of the g.i. tract. It is convenient to consider the inorganic or mineral elements and their compounds separately from organic compounds, which make up a substantial part of any diet and are also frequently encountered as contaminants. The simpler the chemical structure of a substance, the more direct is its ingestion and utilisation within the body, and this applies particularly to the compounds of the mineral elements. Many of the mineral elements will pass readily into solution in the acidic stomach juices and will be quickly transported from the upper part of the g.i. tract into the blood or lymph plasmas and in to the organs and tissues of the body. The chemical affinities of the mineral elements and especially their similarities with elements normally utilised by the body will largely determine the extent to which these foreign substances are concentrated in different tissues and also the speed with which they are eliminated. The relationships between radioactive strontium and calcium and between radioactive caesium and potassium are particularly relevant in this connection, and form an important part of Chapter 4.

There are a few compounds of mineral elements which are insoluble in the stomach juices, and they are usually refractory oxides of metals, such as an oxide of the man-made element plutonium. The contaminant in these cases may be present in food as a small insoluble particle,

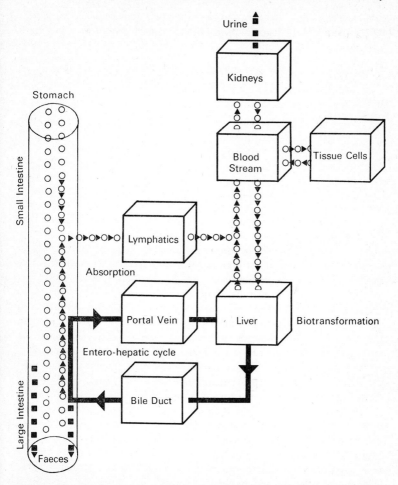

Fig. 1.4 The distribution pathways for substances entering the body via the food consumed. The figure is simplified diagrammatically to represent the events involved in the later stages of the digestion and absorption of food. It should be noted that the bile secretion from the liver or gall bladder, as also the secretion of the pancreas, occur in the upper part of the small intestine to assist in the catabolism, i.e. breakdown of the food substances

which has been trapped superficially on the surface of a growing plant. An insoluble particle should pass through the g.i. tract and be eliminated fairly rapidly in faeces. The g.i. tract is then the critical target for any damaging effects. In the case of plutonium it would be exposed to the radiations emitted during its transit.

The organic pollutants such as a majority of the pesticide chemicals show a considerable range and variation in their chemical composition and structure. They are absorbed at varying rates in passing through the g.i. tract and these rates will often be influenced by their solubility in fatty substances in the food or present in cells lining the intestinal wall. There are a few relatively simple organic substances (acetyl salicylate or aspirin is one example) which are absorbed quite rapidly in the upper part of the digestive tract, but many other substances are only absorbed lower down the tract and after undergoing some form of chemical modification of their composition. This applies especially to the major nutrients in food; proteins are decomposed into the much simpler animo-acids, starches into simple sugars such as glucose, and fats into glycerol and fatty acids. The various forms of organic pollution may not undergo such radical changes in composition before absorption takes place; in many cases the change may be a relatively simple one such as the addition or deletion of a simple grouping such as the hydroxy-group (OH) or the amino-group (NH_2), in the original structure. These and other forms of bio-transformation may continue after absorption has taken place and especially in the tissues of the liver.

Absorption of foreign compounds entering the body as food contaminants, as in the case of essential nutrients, is very largely accomplished in the small intestine of the g.i. tract. After passage through the lining of the small intestine, the majority of substances enter the venous blood and are rapidly transported via the portal vein to the liver. Some absorption directly into arterial blood or lymph may also occur resulting in a more rapid and general or systemic circulation. Transport of a foreign substance frequently involves its binding to proteins or to blood cells. Activation of the substance may then involve its release from binding at the site of action.

The liver is quite a remarkable organ as apart from its important role as a storage depot for certain nutrients it is able to transform many foreign substances and especially organic compounds into forms which have different physico-chemical properties from the original substance and favour their more rapid elimination from the body. This screening action of the liver is associated with certain specialised cells, which contain a variety of non-specific enzymes capable of transforming foreign compounds by the substitution or deletion of simple groups (OH or NH_2). The transformed compounds generally have a greater degree of electrical polarity, rendering them more soluble in water, and assisting their elimination by the kidneys. In this capacity the liver is capable of detoxicating many potentially hazardous foreign substances, but this ability to transform a great variety of foreign substances can sometimes lead to an increase in toxicity or even the potentiation of a normally non-toxic substance. The harmful properties of polycyclic hydrocarbons for instance are believed to be due to the formation of an intermediate oxygenated compound (an epoxide), the original hydro-carbons and their more normal metabolic products, dihydroxy-compounds having a much lower toxicity. Biotransformations take place mainly in the liver although certain changes may also occur in almost any tissue. The changes occurring in the liver are then either advan-tageous to man and other animals when toxic substances are trans-formed into harmless or less toxic substances; they may sometimes be disadvantageous when the reverse occurs and relatively harmless substances are converted into more toxic forms.

Variations in the activity of the liver between different species of animal are a vital factor when attempting to assess human susceptibility to the hazardous effects of foreign substances by screening experiments involving animals. There can be species variations in the response to toxic substances and these can be due to differences in the liver enzyme systems between the species of animals, and hence differences in their ability to bring about biotransformations. These differences are illus-trated by a potential insecticide 2-acetylaminofluorene, which was found to be capable of inducing tumours in the livers of experimental rats. The guinea pig on the other hand was found to be resistant to tumour induction, and this was attributed to the absence of a specific enzyme capable of transforming the original compound into a carcinogenic product. This particular compound might therefore have been cleared for use as an insecticide if the experimental studies had been restricted to guinea pigs. The existence of species variations in the response to foreign substances merely serves to emphasise the difficulties in obtain-ing reliable data from trials with experimental animals and having to extrapolate the results to man.

The original or transformed substance may be stored temporarily in the liver, or it may be partially secreted via the bile duct back into the small intestine, or it may enter the bloodstream. The original or modi-fied substance is generally combined or associated with other substances in the bile juices; once it enters the small intestine, it may be broken down again and either be re-absorbed or eliminated in the faeces. This entero-hepatic circulation may be repeated a number of times with a gradual reduction in the amount circulating due to faecal elimination or entry into systemic circulation. The substance entering the bloodstream is either filtered out and eliminated in the kidneys or finds its way into general circulation along with circulating lymph. In the latter case a variety of tissue cells in the various organs of the body are in contact with plasmas containing the substance, which may then be absorbed into the interior of these cells. Damage to the cell may then result from interference with cell processes described in the next section.

Each of the stages in the absorption and distribution of a foreign substance (Fig. 1.4) involves the passage of the substance through a biological membrane. Every such membrane has a complex structure; in some cases it consists of multiple layers of cells such as in the skin and placenta, and in other cases it consists of single layers of cells as in the lining of the small intestine. Individual cells which are the microscopic living units of any tissue of the body are also surrounded by a membrane and certain structural units within a cell such as the nucleus are also surrounded by membranes separating them from the cell protoplasm. Cell membranes are essentially ordered, but dynamic, structures containing organised layers of proteins and fats and known as lipoproteins. Most membranes in biological organisms have a lipoidal character and transport across such membranes favours substances which are soluble in fats. The membrane may also incorporate fine pores generally lined with water and through which small water soluble molecules may also pass. These pores normally represent a very small fraction of the membrane surface. These types of transport processes are generally regarded as passive, but other mechanisms for the transport of small molecules and ions also exist. These mechanisms involve the expenditure of energy and are regarded as active transport processes. They appear to be largely involved in the transport of essential nutrients such as simple sugars and amino-acids.

Membranes are important for maintaining the structure and the integrity of a cell or organ, and at the same time they must function correctly to allow the influx of essential nutrients and the removal of waste products. The transport of any substance around the body, its accumulation and retention in any specialised tissue and its eventual elimination all involve transport processes across membranes. These processes may be specially important in the kidney where the integrity of the membranes which filter out the waste products is vital for the health of the individual. All the processes involved in the transportation of foreign substances, in any biotransformations they may undergo, in their accumulation and utilisation within any tissues or organs of the body are all time dependent processes. The exposure of any tissue or organ is therefore a complex function of time; generally if the intake from the food consumed is at a steady rate the concentration in the tissue or organ will build up to a steady or equilibrium value over a certain period of time. Once the intake has ceased or after a single exposure to the contaminant the concentration will then decrease with time in accordance with the exponential law which is characterised by a biological half-life.

The biological half-life is the time in which the total amount of substance in the body or its concentration in a special tissue of the body is reduced to one-half of the original value. Biological half-lives can vary considerably from quite short periods of hours up to several tens of years. Substances laid down in fatty tissues or in the mineral bone generally have quite long half-lives, whereas other substances which tend to remain in the body fluids may have quite short values. The biological half-life should not be confused with the radio-active half-life which is the time during which the amount of activity of a radio-active substance is halved. In the case of internal exposure to radio-active materials the pattern of exposure depends on a combination of both the biological and radio-active half-lives.

Where accumulation is taking place over a long period of time, the time which it takes to reach the steady state in the body is also dependent on the half-life; the shorter the half-life the faster the approach to the steady concentration value and vice-versa.

MECHANISMS OF BIOLOGICAL DAMAGE

A GREAT many chemical substances, ionising and other forms of radiation such as ultra-violet, together with a variety of viruses and pathogenic bacteria are all agents of biological damage in man. The damaging effects are essentially twofold:

1. Somatic effects occurring in body tissue exposed to the agent and affecting directly the health of the individual.
2. Genetic (or hereditary) effects occurring in the gonads (the germ cells) and affecting the health of descendants through several generations.

The somatic effects of the pollutants are many and varied; they may include carcinogenesis, i.e. the induction of cancer in a wide range of tissues, damage to nerve cells, mental impairment, and infectious disease. The damaging effects may occur at any stage of growth and development from conception to adult life. Teratogenesis involves damage to the foetus or embryo and results from exposure to the agent at any time after conception; the example of thalidomide is only too well known. In general the effects of exposure are likely to be more severe in early prenatal life when tissue growth is at a maximum rate, and the opportunity for interference with the complex sequence of events leading to cell division and differentiation are obviously greatest. Serious mental damage can also occur in the early years of life. About 80% of the central nervous system (CNS) is developed by the age of 18 months and 90% of the brain structure by the age of 4 years; impairment of the cells comprising the CNS and the brain by early exposure to agents such as lead and mercury may then have profound consequences for the mental development of the individual. Certain agents may also be responsible for an increase in abortions, still births and infant deaths when the exposure has occurred after conception.

The genetic effects of exposure to pollutants, mutagenesis, are by contrast limited to the germ cells of the body. The primitive germ cells of an individual are formed at an early prenatal stage of cell differentiation, and damage to the hereditary material can occur at any stage of an individual's reproductive life. The changes which occur in the germ cells do not affect the health of the exposed person and may only be expressed as hereditary effects passed on to future generations. These changes or mutations range from quite harmless variations such as the colour of the eyes to crippling disabilities such as mongolism, haemophilia, dwarfism and may also be responsible for increased abortions, still-births and infant deaths. There may also be genetic factors resulting in a predisposition to degenerative diseases such as diabetes to metabolic deficiencies, and an increased susceptibility to various ailments.

The potentiality of certain chemicals to induce genetic mutations has been known about thirty years and the effects of ionising radiation even longer. The damage to other tissue cells leading to somatic effects may also be a consequence of genetic changes taking place in the body cells although other mechanisms such as enzyme inactivation and membrane damage may also be involved. All the cells making up the tissues of the body incorporate a number of structures such as the cell nucleus, the protoplasm, which itself has a highly complex structure, and the membranes which contain the cell and divide the nucleus from the protoplasm. The various cells of the body all have the same basic components but are differentiated in various ways to accomplish the specialist functions within the particular tissues in which they are found. The nuclei of the somatic cells hold the coded information which directs and controls the normal functioning of the cell.

Damage to tissues results from damage to the cells which populate the tissue. The harm at the cellular level may be initiated by changes at the molecular level resulting from the interaction of cell components with the pollutant. These changes include:

1. Damage to the cell membrane impairing its normal function of transporting nutrients and other substances to and from the cell.
2. Damage to enzymes within the cell and impairing the normal metabolic processes which support the functional role of the cell.
3. Damage to the nucleus and especially to genetic material; this damage may result in a mutation of the genetic material affecting hereditary in the case of the germ cells or the code of instructions for the proper functioning of the somatic cells.

The nucleus of any cell has a complex structure and the genetic endowment is contained within chromosomes which are revealed by staining techniques when the cell is about to undergo division. Examina-

The Nucleotides are condensation products of :

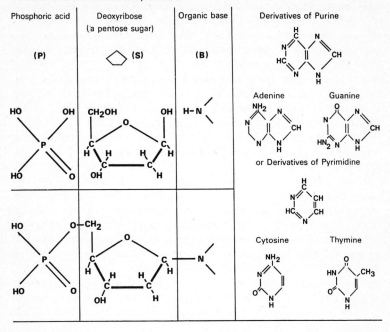

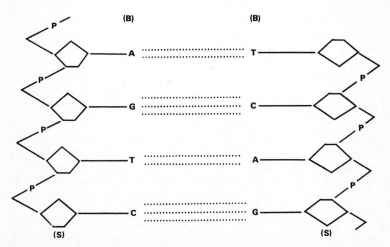

Fig. 1.5 The composition of DNA. DNA consists of a polymeric chain of nucleotides. In the cell chromosome, shown in the lower part of the figure, one polymeric strand of DNA is linked with a complementary chain

tion of a chromosome under the electron microscope reveals a fine structure of fibres in which the principal substances are giant thread-like molecules of DNA (deoxyribonucleic acid) in association with long chain proteins, the histones. Each fibre consists of two complementary molecular chains of DNA, wound in the form of a right-hand helix along the chromosome axis.

The structure of DNA holds the key to cell functioning and is of very special interest in displaying the complexity of some of the vital substances in living matter and of the biological processes in which they are involved. This complexity also emphasises the opportunities which can exist for interference with living processes at the molecular level whenever unusual chemical or physical agents of damage are around.

A single strand of DNA is an example of a natural polymer, a poly-nucleotide, consisting of up to 20,000 simpler molecules the nucleotides, linked end to end. Each nucleotide is also fairly complex, consisting of three simpler types of molecule, viz. phosphoric acid (P), a sugar deoxyribose (S), and an organic base, an aromatic nitrogeneous compound (B). The sugar is a monosaccharide, but with a short chain of five carbon atoms instead of the more usual six as in glucose; in the nucleotide four of these carbon atoms are joined with an oxygen atom to form a five-membered ring, Fig. 1.5. The sugar then forms a bridge between the phosphoric acid at one end and the organic base at the other end. Successive nucleotides are linked in the chain through the phosphate group which with the sugar provides the linear network for the attachment of the sequence of bases. In the case of DNA the base is either a derivative of pyrimidine (thymine (T) or cytosine (C)) or of purine (adenine (A) or guanine (G)). Examples of more familiar purines are caffeine found in tea and coffee and theobromine found in tea and cocoa. The overall structural pattern is also represented in Fig. 1.5. Each strand of the chromosome consists of two complementary chains of DNA making up the double helix and there are additional linkages involving hydrogen atoms between adjacent pairs of bases. These link-ages provide extra rigidity to the chromosome strands and are only possible between complementary pairs of bases either adenine and thy-mine or guanine and cytosine. The unit of information or of heredity, the gene is represented by a minimum sequence of three bases along a short length of the chain. The total number of genes associated with the human cell's complement of twenty-three pairs of chromosomes is estimated to be about 6000 M.

The whole body is made up of about 10^{15} cells all differentiated from the original fertilised egg, to which each parent has contributed half of the chromosomes and genes. Each differentiated cell contains the same pattern of chromosomes and genes, although not all genes will be expressed in the particular tissues in which the cell is to be found. Every cell is a dynamic entity constantly building the variety of proteins essential for its own proper functioning and that of the whole organism. A cell eventually divides to form a pair of daughter cells when its normal life-span is completed. The lifetimes of cells vary considerably; those which line the walls of the gastro-intestinal tract and the stem cells of the bone marrow, for instance, divide rapidly, whereas the cells of the brain and nerve tissues are almost permanent. In the case of somatic cells exact replication of all twenty-three pairs of chromosomes in the daughter cells occurs. Just prior to division the two complementary chains of DNA in each strand of a chromosome separate; each separate chain then serves as a template for the recreation of its exact complementary partner, which is achieved with the assistance of a special enzyme within the cell, DNA-polymerase. This ensures the identical sequence of nucleotides in each chromosome strand of the divided cells and hence an identity of function. In the case of the germ cells the pattern of division is the same until the final division leading to mature germ cells; at this final stage duplication of the chromosomes does not occur so that each mature germ cell carries only one half of the full complement of 46 and ensures variation in the offspring in which features and qualities of either parent may be expressed.

A vital process of most cells is the synthesis of a number of essential proteins. All proteins are further examples of natural polymers, consisting of long chains of varying combinations of about 20 amino-acids. The quality of a protein and its functioning in the body, depend on the number and correct sequence of the amino-acids and also on the length and the shape of the chain. The complicated synthesis of the cell proteins is controlled by the information encoded within the DNA, but is mediated by other polynucleotides, messenger and transfer RNA and enzymes within the ribosome bodies of the cell protoplasm. RNA (ribosenucleic acid) is very similar to DNA, but has ribose instead of deoxyribose as the sugar and one of the pyrimidine bases is replaced by uracil.

The proper functioning of cells clearly depends on the exact replication of the molecular processes involved in protein synthesis and cell division. These processes must also be dependent on the availability at the right time of the right metabolic substances obtained from the food consumed. Foreign substances in the food, and physical agents such as ionising radiation can either hinder the normal processes or interact with cell components and their normal functions within the cell. Many chemical carcinogens or their active metabolites appear to be characterised by a strong electrophilic character (electron attracting). These compounds are therefore able to interact with nucleophilic centres (electron donors) in the nucleic acids – the purine or pyrimidine bases.

The resultant damage to a base may alter the normal pairing of the bases in the double strand causing a point mutation or it may lead to a chromosome break by rupture of the DNA chain. The precise sequence of events is not known with certainty but this type of interaction appears to be an initiating stage. It is a rapid and generally irreversible step, but may be followed by a long period of promotion when the adverse change becomes apparent in the organism.

All forms of cancer exhibit this latent period between the initial step and the actual appearance of the cancer, which in some cases may be many years later. Other agents, e.g. certain viruses, may be involved in the promotion of a cancer and the combination of two or more agents may be necessary for a full induction of a cancer. The carcinogenic properties of certain chemicals have been known since 1895 when cases of bladder cancer were associated with 2-naphthylamine. Many examples of other highly carcinogenic chemicals are now known; they include polycyclic aromatic hydrocarbons such as the well-known benzopyrene (present in cigarette smoke, diesel fumes, soot and tar) mustard oils and nitrosamines (or their precursor nitrous acid). They may also include some naturally occurring chemicals such as the aflatoxins, produced by fungal growths on groundnuts. Some of these chemicals are also potential contaminants of food. Despite the known carcinogenic properties of many chemicals, protective measures in industry and elsewhere have generally failed to match the scale of precautions taken with ionising radiation.

Any biological damage as the end result of damage to cells can only occur if the cell remains viable, continuing to divide and to function in a deleterious manner. A cell which is severely damaged may no longer remain viable and is then just eliminated from the tissue without any resultant effect. The body's own immunological defences may also offer

Table 1.6

Examples of natural mutations

| Mutation | | Effect | Estimated |
Type	Cells		Natural frequency
Point	Genetic	Achondroplasia (dwarfism)	$10/10^6$ germ cells
		Haemophilia (bleeder's disease)	$30/10^6$ germ cells
		Phenylketonuria (PKU)	1/20,000 births
Chromosome	Genetic	Down's syndrome (Mongolism)	1/700 live births
		Klinefelter's Syndrome (spontaneous abortions)	1/150 live births
	Somatic	Leukaemia	$40/10^6$ population pa

additional protection, the mutated cell functioning as an antigen with the formation of antibodies leading to its destruction. Multiple exposures to a great variety of pollutants throughout a lifetime, and the natural ageing of the individual may, however, be important factors in breaching the body's own defences.

The mutations occurring in somatic or genetic cells are basically of two types – gene (or point) mutations involving a localised alteration in the DNA double helix, and chromosome mutations which are microscopic changes in the chromosomes which can be made visible when cell division is occurring. The latter are also two-fold in nature – changes in the actual number of chromosomes (genome mutation or aneuploidy) or aberrations involving reorganisation of chromosome fragments after a break has occurred. Some examples of the various types of mutation with some estimates of spontaneous occurrence rates are listed in Table 1.6.

BIOLOGICAL EFFECTS

THE MECHANISMS leading to damage at the molecular and cellular levels in the body tissues with reference to the subsequent effects in the tissues of the organism have been considered in the previous section. Chronic long-term effects resulting from internal exposure to a variety of polluting agents are the principal concern of this book and an assessment of the consequences for the future health of the population is a matter of vital importance. The nature of these effects are summarised in this section and are subsequently related to the exposures from various contaminants in the later chapters. In addition to the chronic effects of certain agents on brain and nerve tissues with the risk of impaired mental health, the principal effects are as follows:

1. *Carcinogenesis* or cancer induction. The process of induction may involve a mutation in somatic cells exposed to the contaminant, but other factors may also be involved in promoting the actual tumour. Cancer induction may be a consequence of exposures occurring at any stage in an individual's life, including the pre-natal stage with the risk generally increasing with the concentration of the contaminant and the duration of the exposure.

Cancer has now become a major source of deaths in the developed Western nations, rivalling the incidence of deaths due to cardiovascular disease. It is still the subject of much current medical research, although it is still not clear whether a carcinogen can initiate the development of a cancer directly, or whether it acts by hastening the processes which lead to spontaneous occurrence. It seems fairly clear that the interplay of several factors including a number of viruses, numerous chemicals and ionising radiations may all be involved. The various agents of internal pollution must be considered as additional factors involved in the general increase in incidence of cancer.

2. *Mutagenesis* or damage to the hereditary material of the germ cells which may occur at any time during the reproductive lifetime of the individual. The damage is expressed in future generations as physical or mental abnormalities or as changes in certain physiological functions. Genetic changes may also predispose towards chronic ailments such as cancer and to a general lowering of the health of the population.

Mutagenesis, then, involves harm to future generations without affecting the health of the exposed individual. Those mutations involving chromosome aberrations are generally quite drastic and are expressed in the generation immediately following the mutation. At their most severe, they can cause embryonic or foetal damage or be responsible for severe mental and physical disabilities. The gene or point mutations, on the other hand, are normally recessive and may be hidden in the genetic pool of the population for several generations until an identical mutant is encountered in the conjunction of the sperm and egg cells. The effects of gene mutations can vary appreciably in their severity, from genetic deaths and genetic diseases, involving about one per cent of all live births, to quite mild ailments with certain types of metabolic disorder. There is some evidence that mild mutations are far more frequent than the severe ones and there has also been a suggestion that recessive gene mutations may even be partially expressed in earlier generations, resulting in a more rapid general decline in the health of the population.

3. *Teratogenesis* or damage to the embryo occurring during pregnancy through the exposure of the mother and passage of the pollutant through the placenta. The resultant defects may include the all too familiar gross physical abnormalities, but physiological and possibly mental damage can also occur.

Teratogenesis is probably the result of mutations in the body cells of the rapidly developing embryo, which is especially sensitive to the effects of an exposure in the early period of gestation when organogenesis is extremely rapid. In the case of the human embryo, this period is from the end of the first week to the twelfth week of pregnancy.

A common factor may well be involved in all these three effects, namely mutations taking place either in the body cells or germ cells. Any agent which is suspected of causing any one of these effects may well, therefore, be a possible factor in all three.

4. *Enzyme interactions* involve the stimulation or inhibition of the active enzyme systems in the liver. These systems normally moderate the toxic effects of foreign substances in the body, with the result that modification of the activity of the liver through the effects of one agent may alter its ability to deal with a second agent.

Foreign substances, among which food additives and medical drugs must also be included with the environmental pollutants, may interact in several ways with the tissues of the body. One of the most important of these interactions is the one which occurs at metabolic sites in certain cells of the liver. The versatility of this organ in dealing with foreign substances has already been described. In certain cases, a foreign substance may stimulate the activity of the liver enzymes and, in other cases, it may act oppositely, inhibiting enzyme activity. At present, there appears to be no method of forecasting whether a particular chemical will be a stimulant or an inhibitor. DDT and other chlorinated hydrocarbons are known, however, to be potent stimulators of enzyme activity and may well, therefore, interfere with the action of certain medical drugs. A drug which is normally activated or potentiated by the action of the liver, by its transformation into an active metabolite, will have its potency enhanced, although it may act for a shorter time. On the other hand, a drug which is administered in active form may be deactivated by enzyme interactions and will, therefore, have its potency considerably reduced. There is also a possibility that enzyme interactions, involving pairs of foreign substances, may also give rise to synergistic effects in man, the action of two substances together being greater than the sum total of their individual effects. With the large number and variety of foreign substances entering the body, this poses many serious questions for which there is no satisfactory answer at the moment. It has been suggested that "the marked individual variations in the metabolism of foreign substances in man raises the possibility that the presence of inducers and inhibitors in food may already have caused marked changes in the ability of the human population to eliminate foreign compounds".[7]

NATURAL AND ARTIFICIAL EFFECTS

THE SPONTANEOUS occurrence in man of most of the effects which may be caused by a variety of pollutants complicates enormously the problems of analysing and identifying effects due to pollutants. In the case of man, most of the evidence is indirect and based on experiments with other organisms such as the fruit fly or mouse and other small animals or on experiments with cell cultures in vitro. Usually, in order to obtain significant and statistically reliable data the experiments have to be conducted at exposures to the pollutant well above the present levels of human contamination. The evidence for effects on man must then be estimated on the basis of doubtful extrapolations to low dose levels and from a much lower order of species to man. In very many cases the evidence can only be regarded as presumptive for man and precautions must then be based on the assumption that proportional damage occurs at much lower doses.

The spontaneous occurrence of most effects and the possibility of higher rates of incidence (due to perfectly natural causes) is a further difficulty in attributing environmental effects to man-made pollutants. Dr Perkins has recently drawn attention to this problem in relation to the simultaneous occurrence of three fish diseases in the Irish Sea.[8] They appear to have affected the fish, dabs, in epidemic proportions during 1971, and all three have occurred in separate outbreaks on earlier occasions. The three disease conditions are limphocysti (an unsightly swelling of skin cells), ulceration of the skin, and fin damage. It is tempting to attribute the simultaneous occurrence of the three diseases to the large-scale dumping of a great variety of human and industrial wastes into the middle of the north Irish Sea, but proof of this relationship is an extremely difficult matter. It is just possible that all three outbreaks could have flared up together. The results may point to another possibility that the presence of low concentrations of certain pollutants in the sea-water may be lowering the resistance of organisms and hence their ability to combat viral infections in their normal habitat.

Whilst direct circumstantial evidence linking cause and effect may be lacking in the case of man, there are obviously grounds for serious concern. There is an imperative need for an expansion of experimental work to obtain relevant data to enable a reliable evaluation of each situation involving one or more polluting agents before a final verdict can be reached. In the meantime, all attempts to clean up the environment and to observe all reasonable precautions affecting the quality of food are obviously to be welcomed.

RISK RATES

FROM THE moment of birth, human life is fraught with a great variety of risks. Many of these are due to natural causes such as fire, plague, earthquakes, volcanic activity, wild animals and meteorological phenomena. The effects of some of these natural causes can be mitigated by man-devised efforts; additionally man has also introduced a great variety of unnatural risks, which in some cases are now an accepted part of civilised life. Whilst the risks attendant upon the human urge for adventure are much reduced or at least restricted to much smaller numbers of people, the demands for more and more rapid forms of communication and the products of modern technology have affected increasing numbers of people. Modern living has not succeeded in abolishing the risk to life and limb and in some cases has increased it appreciably. Estimates of risk rates from a variety of occupations and human activities are listed in Table 1.7.

Table 1.7

Relative risk rates (after Sowby)[9]

	Risk	Rate no/hr in 10^6 population
Occupational	All males	0·1
	Chemical industries	0·034
	Fishermen	0·33
	Coal miners	0·40
	Railwaymen	0·45
	Construction engineers	0·675
	Steeplejacking	3·0
Non-occupational	At home, total	0·026
	At home, electrical	0·001
	At home, old folks	0·013
	At home, children under 6	0·003
	Lightning	0·0001
	Road travel by car	0·57
	Road travel by motor cycle	6·6
	Cigarette smoking	1·2
	Air-flying	2·4

High risk rates involving dangerous occupations or activities are generally limited to small numbers of people and do not contribute significantly therefore to the total number of annual deaths in the human population.

There is a tendency to accept certain risks as the price for some of the benefits of modern living, and in some cases the acceptance involves almost a sense of complacency or even indifference. In these cases there is evidence for a more or less conscious balancing of the risk against the benefit. It may become equally necessary to strike a balance in the case of certain pollutants in food, but this balance must be based on an appraisal of the risks and the benefits in so far as this is possible with the available data.

A NOTE ON UNITS

THE BASIC SI units of length and mass are the metre (m) and the kilogram (kg) respectively.

1 m = 3·28 ft
1 kg = 35 oz

(See *Metrication in Scientific Journals*, The Royal Society Conference of Editors, 1968.)

Fractions and multiples which are recommended for use with these units are as follows:

Fractions		*Multiples*	
milli (m) $= 10^{-3}$		kilo (k) $= 10^3$	
micro $(\mu) = 10^{-6}$		mega (M) $= 10^6$	
pico (p) $= 10^{-12}$		giga (G) $= 10^9$	
		tera (T) $= 10^{12}$	

Other units which involve the units of length and mass and which are also allowed to be used in conjunction with SI units are as follows:

Mass: tonne (t), $1 \text{ t} = 10^3 \text{ kg} = 1 \text{ Mg}$
Area: hectare (ha), $1 \text{ ha} = 10^4 \text{ m}^2$
Volume: litre (l), $1 \text{ l} = 10^{-3} \text{ m}^3$
1 tonne $= 2200$ lb, 1 hectare $= 2{\cdot}47$ acres, 1 lire $= 1000$ ml $=$
1·76 pints

The units of concentration by weight are of very special importance in the chapters which follow. When dealing with environmental pollutants, these concentrations are generally very small and are expressed as parts per million (ppm) or parts per billion (ppb). 1 ppm $=$ 1 mg of pollutant/kg of material $= 1\mu$g pollutant/g of material. 1 ppb $= 1 \mu$g pollutant/kg of material. In the case of concentrations in water, density of the solution is assumed to be 1, and hence, 1 ppm $=$ 1 mg/litre and 1 ppb $= 1 \mu$g/litre. The special units of radio-activity and radiation are presented at the end of Chapter 4 (pp. 153-4).

REFERENCES

[1] *The Human Environment : the British view*, prepared on the occasion of the UN Conference on the Human Environment, Stockholm, 1972; published for the Department of the Environment by HMSO, London, 1972, p. 9.

[2] *Pollution : nuisance or nemesis?*, a report on the Control of Pollution, published for the Department of the Environment by HMSO, London, 1972.

[3] *The Threat of World Pollution*, Sir Kenneth Mellanby, Essex Hall Lecture, 1971, The Lindsey Press, London, 1971.

[4] *European Standards for Drinking Water*, 2nd edition, 1970, WHO Regional Office for Europe, Copenhagen, 1970 (EURO 0664).

[5] *Food Resources, Conventional and Novel*, N. W. Pirie, Penguin, Harmondsworth, 1969.

[6] *Household Food Consumption and Expenditure, 1969*, Annual Report of the National Food Survey Committee, Ministry of Agriculture, Fisheries and Food, HMSO, London, 1971.

[7] "Enzyme Induction in Laboratory Animals and its Relevance to Food

Additive Investigations", Chapter 11, p. 276, in *Metabolic Aspects of Food Safety*, ed. Francis J. C. Roe, Blackwell Scientific Publications, Oxford and Edinburgh, 1970. (Includes chapters on the physiological processes of absorption through the g.i. tract, metabolic pathways, and the effects of food additives on the liver.)

[8] "Incidence of Epidermal Lesions in Fish of the north-east Irish Sea", E. J. Perkins, *Nature*, 1972, *238*, 101.

[9] "Radiation and Other Risks", F. D. Sowby, *Health Physics*, 1965, **11**, 879.

Suggestions for further reading

Fundamentals of Ecology, E. P. Odum, Saunders, Philadelphia, 2nd edition, 1959 (useful discussions on the flow of essential nutrients, e.g. energy, protein, in an ecosystem).

Two text books on molecular biology:

The Chemistry of Living Cells, Helen R. Downes, 2nd edition, Longman, Harlow, 1963 (useful on metabolism of foods).
The Molecular Basis of Life, readings from *Scientific American*, introduced by R. H. Haynes and P. C. Hanawall, W. H. Freeman, London, 1968 (useful surveys on the nucleic acids, the expression of genetic information, and the regulation of cell activity).

Drugs of Abuse, ed. S. S. Epstein, MIT Press, Massachusetts, 1971 (useful notes on the long-term carcinogenic, mutagenic and teratogenic effects of chemicals).

"Public Health Hazards from Environmental Chemical Carcinogens, Mutagens, and Teratogens", W. C. Hueper, *Health Physics*, 1971, **21**, 689.

Genetic Load: its biological and conceptual aspects, Bruce Wallace, Prentice-Hall, Englewood Cliffs, NJ, 1970.

2

Agricultural Pollution of Food
Part 1. Pesticide Residues

INTRODUCTION

AGRICULTURE AS the major producer of human food also provides the principal sources of intake for a majority of the pollutants in man. The contaminations of agricultural produce may arise accidentally through industrial operations close to cultivation areas, or on a global scale from, for example, atomic weapons fall-out, or it may arise directly and to an appreciable extent through the application of a wide range of chemicals in modern methods of crop production and animal stock rearing. The contamination of agricultural produce from industrial activities and atomic weapons form the subjects of later chapters, whereas this chapter and the succeeding one will be concerned with the pollution of food arising from purely agricultural operations. The various forms which this pollution takes will be considered under the following headings:

1. Chemicals used in crop protection
2. Chemicals used as fertilisers
3. Antibiotics and hormones used in intensive stock rearing
4. Pathogenic bacteria in animal produce and wastes
5. Polycyclic hydrocarbons in diesel engine fumes

Crop protection by means of pesticide chemicals leading to pesticide residues in the food prepared for human consumption is the subject examined in some detail in the present chapter. Certain inorganic pesticide chemicals such as compounds of mercury and arsenic are dealt with more fully in Chapter 5. Items 2 to 5 form the subjects of Chapter 3.

The contamination problems involving the more persistent pesticide chemicals such as DDT may be expected to show a gradual improvement as several countries have recently imposed a ban on certain applications of these chemicals. This should in the course of time lead to a reduction in the contamination levels of the more important food sources, and hence in human contamination, although the chemicals will be around for some time ahead. They also provide an object lesson of the sort of problems that are generated when the full environmental consequences of any action are not thoroughly investigated and evaluated beforehand. Public awareness of these problems was only fully alerted with the publication of Rachel Carson's book, *Silent Spring* in 1962.

THE NEED FOR CROP PROTECTION

SUCCESSFUL AGRICULTURE is the base on which all human existence depends. Its importance for meeting the increasing food demands of the growing population and for improving the standards of nutrition in the developing countries is undeniable. It is not surprising that with the greater pressure on the land to improve both the quantity and the quality of the produce traditional agriculture is being subjected to radical and in some cases revolutionary changes. Traditional agriculture has normally followed a regular system of crop rotation with the cultivation of several arable crops alternating with periods of pasture and livestock. An illustration of such a pattern of rotation (and a rather long one) is the one used at the Haughley Research Farms Ltd, on the organic and mixed (i.e. organic plus chemical fertilisers) sections of the farms.[1]

Year	Crop
1	Wheat
2	Kale and maize
3	Barley
4	Beans/peas
5	Oats
6	Oat silage
7	Ley
8	Ley
9	Ley
10	Ley

The systems of rotation traditionally followed are being replaced by intensive systems of cultivation or stocking, when an identical crop is grown regularly on the same ground. Intensive cultivation or single cropping provides greater opportunities for the establishment of a variety of pests which attack the crop, and increases the need for alternative methods of control. The green revolution which is making a major contribution towards solving the extremely difficult and urgent food problems of the developing nations of the world by introducing new high-yield varieties of staple cereals such as wheat and rice may also increase the opportunities for the establishment of a variety of pests, The need for further developments producing highly resistant strains of the new grains and the risks of depending on a few high yield varieties are fortunately fully appreciated. Transportation of foodstuffs on an international scale also enhances the opportunities for the spread of pests and of their establishment in unusual surroundings.

All agriculture represents a departure from the natural conditions which prevail in the absence of human intervention. All plants in their natural conditions are subject to attack by a variety of pests ranging from fungi, viruses, insects and animals; these pests have continued to exist throughout the traditional development of agriculture and have always been a great harassment to man both during cultivation and storage. At their most spectacular they are represented by the great plagues of Egypt, almost certainly due to swarms of locusts, the outbreaks of ergotism (St Anthony's fire) in the 11th and 12th centuries due to fungal attacks on harvested rye, and the 19th century potato blights in Ireland, again due to a fungus, Phytophthora infestatans; the potato blights were responsible for widespread starvation with many deaths and led to large scale emigration. The consequences of outbreaks similar to the latter today and especially if they should affect the new high-yield grains, could obviously be extremely serious. Epidemics of plant disease such as yellow rust on barley, fire blight on peas, do occur in the UK from time to time although not on the scale of the potato blights. Apart from the losses due to viral, fungal and insect attacks on plants cultivated crops are also in competition with other prolifically growing plants, the weeds. The yield of a crop can be seriously affected if weeds are allowed to flourish so that their control by herbicides is also important in modern agriculture.

THE DEVELOPMENT OF PROTECTIVE CHEMICALS

MAN HAS always been aware of the losses which can occur either during the cultivation of a crop or during its storage, and has throughout recorded history attempted from time to time to exercise some control by the application of chemical protective agents. Arsenic compounds are reported by Plinius to have been used as insecticides as early as AD 70 and arsenicals were also widely used in China in the 16th century. Compounds of inorganic elements such as copper (Cu) and mercury (Hg) were also employed in Europe in the 19th century as fungicides. Bordeaux mixture for example is formulated by the addition of milk of lime to a solution of copper sulphate. This mixture was developed and applied originally to vines adjoining public pathways to discourage passers-by from picking the grapes. One year during a serious outbreak of mildew it was found that vines which had been so treated were immune from the fungus and thriving, so that the mixture was then developed and applied to the protection of hops and vines against fungal attack. The active agent in the mixture is copper which kills fungal spores at quite low concentrations. A suspension of lead arsenate was developed somewhat later for the control of the larvae of the gypsy moth. Mercury and copper compounds are still used today for dressing grain seeds to protect the plant from fungal attack, and arsenic compounds for the protection of orchards.

The swing away from inorganic pesticides began seriously after the end of the Second World War. DDT, a member of the class of compounds known as the chlorinated hydrocarbons, was discovered by a Swiss chemist in the 1930s and was first used extensively as a hygienic insecticide during the war years, for the control of lice as typhus carriers. It was subsequently applied to the control of the malaria carrying mosquito and with such spectacular success that the average expectancy of life in parts of India had been almost doubled after about twenty years of application. This increase in the average life expectancy in India and other developing countries has widened the gap between the birth rate and death rate and in conjunction with other medical advances in these countries has been a powerful factor in the rapid growth of population.

It has been estimated that "more than 1000M people are now living in areas that have been freed from endemic malaria".[2a] The chemical can be applied at two stages of the insect's life cycle – the spraying of surface waters to control the larvae of the insect and the spraying of the home to control the mature mosquito. This eradication of malaria from large areas of the world can be ranked as one of the outstanding achievements of the applications of chemicals to control the so-called vector-borne diseases of man. In addition to insects these may also be transmitted by molluscs such as the snails carrying filariasis and rodents carrying a variety of diseases to man. The World Health Organisation undertakes an intensive search for safe chemicals to be used as insecticides, molluscicides, or rodenticides in the control of public health. Although DDT has been used very largely in the past for many purposes and has proved effective, a number of factors such as its persistence in the environment and the acquirement of resistance to it by various pests are turning the search in other directions. This search for a number of agricultural chemicals which will be rapidly degraded in the environment, and as safe as possible in practice, is turning attention to a number of derivatives of carbamic acid and also to several of the class known as organophosphates.

Many of these pesticides, such as the organophosphorus compounds first applied in agriculture in 1948 and derived from war time nerve gases, have been developed and used susbsequently to the application of DDT. The search continues for more and more specific chemicals which will be destructive of a particular pest without causing the indiscriminate damage sometimes affecting beneficial organisms which is typical of the wide spectrum materials such as DDT. Of all the chemicals which have been produced, however, DDT probably represents the greatest potential hazard to man as a pollutant on a global scale. This results from its chemical stability and persistence both in the environment and in the organism, its affinity for fatty tissues, and its manufacture on a massive scale for applications in agriculture, human disease control, and in the home. DDT will therefore claim a major share of attention in this chapter with references to other pesticide chemicals where appropriate.

The polychlorbiphenyls (PCBs) have a number of features, such as their chemical composition and their persistence, which they share with the chlorinated hydrocarbons widely used in agriculture. They were already being manufactured and applied on a large scale in the early 1930s and their presence in the environment must date from this time as a result of waste discharges. This presence has, however, only been detected in recent years through the development of precision methods of analysis for the chlorinated hydrocarbons, with which they have undoubtedly been confused in the past. They are oily liquids of good electrical properties and have been used in the formulation of plastics, paints, and varnishes, in transformers and as lubricants. They are reported to extend the active life of insecticide formulations although they do not appear to have been used commercially for this purpose. In contrast with DDT and other pesticide chemicals which can only enter man as a result of their movement through food chains, PCBs may also enter foods more directly as they are used in the formulation of some plastics used in packaging materials. It seems appropriate therefore to consider PCB levels in food and in man in this chapter.

There is also another and altogether simpler group of compounds, the halogenated aliphatic hydrocarbons, which include a number of volatile materials such as methyl bromide, CH_3Br, dibromomethane, CH_2Br_2, ethylene dibromide, $C_2H_4Br_2$, and carbon tetrachloride, CCl_4. Several crops are treated with a compound of this class either prior to loading on to a ship or in the ship's hold; the crops include cereal grains, pulses, and nuts. These compounds are valuable in protecting the crop because of their degree of penetration throughout the bulk and their high toxicity to insects. The residues on the crop are either the unchanged compound adsorbed on to the surface or bromide ion due to chemical interaction. The latter is the more likely form of contamination and values as high as 200 ppm, compared with a natural bromide level of a few ppm, have occasionally been found on crops imported into the UK. The toxicological data on the original compounds or on the bromide is limited although tolerance levels have been recommended by the World Health Organisation.

Carbon tetrachloride which is sometimes used as a crop fumigant is also an important industrial product with an annual production of about 1 Mton mainly for conversion into chlorofluorocarbons such as CCl_2F_2, CCl_3F, now widely used as aerosol propellants and refrigerants. Several other compounds are also used as dry cleaning agents and in the formation of PVC. All these substances have now been detected in the atmosphere and oceans of the world but the details of their distribution and their significance in the marine environment are at present uncertain although they are unlikely to be of any great importance.

CHEMICAL PESTICIDES

THE PROTECTION of crops in the field (or in storage) can be considered under three main headings:

1. Herbicides for the control of a variety of weeds which compete for the available nutrients in the soil and reduce the yield of the cultivated crop.

2. Fungicides for the control of fungi which are harmful to the growing plant and again reduce productivity mainly in the field; a further form of fungal attack during storage of certain crops under moist conditions can lead to the production of highly toxic agents in the crop, e.g. ergotine in rye and aflatoxins in groundnuts. In these cases storage under dry conditions also provides good protection.

3. Insecticides for the control of insect pests which can be highly destructive of growing crops and seriously deplete productivity both in the field and in storage.

The importance of these chemicals to agriculture can be judged by their sales in Britain in 1966; herbicides accounted for about £15M of sales, insecticides £7M and fungicides about £3·5M[2b]. A fourth type of crop protection, especially during storage, involves the application of chemicals as rodenticides; the best known example is the substance warfarin for controlling rat populations. Problems of food contamination should not arise from the latter, however, with normal precautions in its application as a bait for rats. Rodent control can also be considerably improved by careful attention to the design and construction of buildings for storage, as the authorities in India have demonstrated.

A very great many insects and types of fungus attack the staple crops such as the cereals on which adequate human nutrition is so dependent. It has been known for a long time, however, that certain insects tend to concentrate their attacks on certain plants and that most plants are generally susceptible to just a few fungi out of the very great variety which exist naturally. There is some evidence that the resistance of certain plants to attack by certain pests may be due to protective chemicals produced naturally within the plant. It is not surprising, therefore, that some of the earliest insecticides to be applied in agriculture were concentrates extracted from special plants. Derris powder is obtained by grinding the roots of the plant species derris and was patented for agricultural use in 1912. The insecticidal properties of the derris plant and also its high toxicity for fish were known to the ancient Chinese. The powder from the roots contains an active ingredient known as rotenone which has a low persistence in the environment. Nicotine is an alkaloid extract from tobacco waste and has been used for over 50 years as an insecticide mainly against aphids. Pyrethrum powder is the ground flower heads of Chrysanthemum cineraefolium (the Dalmation Insect Flower) and has been employed as a domestic insecticide for over 100 years. The active ingredients are compounds known as pyrethrins and cinerins.

The application of various mineral compounds either in their natural state or in a suitably modified form is also a long standing practice. Examples include petroleum oil fractions used as insecticides, tar oil distillates as fungicides, and inorganic compounds of copper, lead, mercury and sulphur used occasionally as insecticides but usually as fungicides. The number of these compounds has only increased slightly in recent years but several organic compounds of mercury have also been developed as fungicides. Other mineral elements such as zinc and manganese have also been incorporated in a range of organic fungicides based on dithiocarbamic acid.

Residues which are due to a mineral or inorganic element as part of a pesticide chemical will certainly have a persistence in the natural environment, and may well lead to some disturbance in the natural distribution of these elements in soils, crops, and in the aquatic environments. These disturbances will also have their effects on man, but as they are further aggravated by industrial waste discharges their discussion is deferred until Chapter 5.

Although there has been some development then in the use of heavy metals, notably copper and mercury by their incorporation in various organic compounds, the major developments in agricultural chemicals since the Second World War have involved an increasing number of purely organic compounds. These can be grouped into a number of important organic chemical classes of which the more important are the chlorinated hydrocarbons, the chlorophenoxy-acids, organophosphorus compounds, the dithiocarbamates, bridged diphenyl compounds, triazines and bipyridyls.[3,4]

The cholorinated hydrocarbons include about twenty compounds all of which have been incorporated at some time in various commercial pesticide formulations. Many of them are distinguished by high toxicity to insects, chemical stability, low biodegradability, and a slight volatility. They are examples of compounds absent from the normal environment and their stability ensures their persistence for some years after any application. Their slight volatility and absorption on to small dust particles are factors leading to their global diffusion and the contamination of rain or snow precipitated in all parts of the world. It has also been shown that there are delays in the movement of these compounds from one part of the environment into another; there is for example a delay of several years in the transfer of DDT from soil to sea-water, so that the maximum concentration in sea-water as a result of DDT applications to crops will only be reached some years later. The best known examples of these compounds are DDT (50% chlorine), aldrin (58% Cl), dieldrin (56% Cl) and BHC compounds (78% Cl). Other compounds having some similarity of chemical structure with aldrin include endrin (56% Cl), chlordane (69% Cl) and endosulfan (52% Cl).

All the chlorinated hydrocarbons are distinguished by a high content of chlorine, part of which is always bound to a ring of carbon atoms. The hexagonal ring of six carbon atoms is found in benzene, which is the simplest compound of a large series of aromatic compounds. Two of these rings are bridged by another atom of carbon in the case of the DDT compounds; aldrin and its related compounds however have a more complicated diene ring system in their structure. The stability and persistence of these compounds in the environment is due to the strength of the bonding between the chlorine atoms and the ring carbons. Compounds with similar bonds between chlorine and an aromatic ring do not occur in significant amounts in the natural environment. The formulae of the most important examples involved in food contamination are shown in Fig. 2.1 (1, 2, 3).

The PCBs are also chlorinated hydrocarbons, consisting usually of mixtures of the chlorination products of naphthalene and other aromatic hybrocarbons. They are distinguished by even greater stability and persistence in the environment, Fig. 2.1 (4).

There are a number of other compounds with two benzene rings in their structure, similar to DDT and related compounds but with other atoms such as sulphur replacing the carbon atom in the bridge. Examples of this group include fenson (5) and chlorfenson (6) which are useful insecticides for controlling ticks and mites in orchards and in glasshouses (*see* Fig. 2.1).

The organophosphates are an important group of insecticide chemicals of more recent development and application than the chlorinated hydrocarbons. They are essentially organic compounds of the mineral acid,

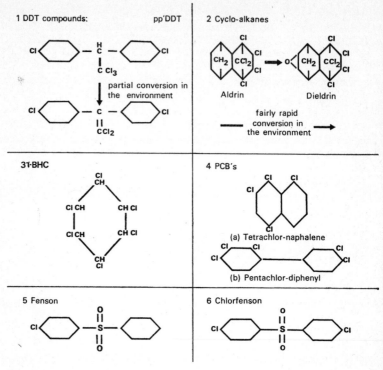

Fig. 2.1 Examples of chlorinated hydrocarbons

phosphoric acid, with in some cases the oxygen atoms of the acid being replaced by atoms of sulphur. They have powerful insecticide properties. Interest in these compounds derived from their war-time development as nerve gases and their ability to paralyse the nervous system. They are usually broken down quite rapidly in the environment and are nothing like as persistent as most of the chlorinated hydrocarbons. On the other hand the mechanism of their toxicity for insects, the blocking of nerve impulses, makes them potentially highly toxic to mammals even in very small amounts. In some cases, however, the toxic agent is a metabolite of the original compound; the selective toxicity of malathion for example against insects is due to an enzyme oxidative process in the insect, whereas in mammals a different enzyme reaction, hydrolysis, occurs before oxidation can take place. Selected examples of these compounds as derivatives of orthophosphoric acid or of related sulphur compounds are shown in Fig. 2.2 (1–4).

Phosphoric acid forms numerous salts with inorganic elements some of which are essential as sources of phosphorus in the soil, and are

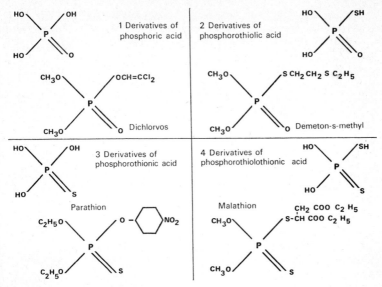

Fig. 2.2 Organophosophates

applied as fertilisers such as superphosphate, and bone-meal or bone-ash, all of which are forms of calcium phosphate. It is also an essential element in the human body and in the diet; it is present in the mineral bone, also in the form of a calcium phosphate, apatite, and the acid itself is an important constituent of nuclear DNA and RNA (*see* p. 25) and also of certain fats and proteins.

When, however, the phosphoric acid is combined with other organic molecules to form esters, the organic equivalents of the inorganic salts, it is completely transformed into compounds which are highly toxic to many insects and mammals.

The sulphur derivatives of phosphoric acid may not themselves be active as pesticides and are only activated by enzyme reactions taking place within the organism. These reactions generally involve the replacement of one or more of the sulphur atoms by oxygen in the phosphoric acid portion of the molecule. These reactions may take place within the insect and may or may not be identical with reactions in an animal. The organophosphates are generally decomposed fairly readily and as a consequence do not have the persistence in the environment of the chlorinated hydrocarbons.

The compounds which have been described so far have all been developed for the primary purpose of controlling insect pests. Further groups have been developed as herbicides to control the growth of weeds competing with cultivated crops. They include a large number of compounds which are derived from chlorophenoxy-acids, bipyridyls, and triazine.

The chlorophenoxy acids (auxin herbicides), the "hormone" type weedkillers, are all based on the structure

with chlorine substitution in the aromatic nucleus and the acid portion being derived from acetic (R = H), propionic (R = CH_3), or butyric acid (R = C_2H_5). They became available in 1948 and are widely used as herbicides; the best known examples because of their large-scale production and use by the American armed forces in Vietnam to defoliate trees and to destroy jungle terrain being:

Other well-known examples of this class are NCPA, MCPB and 2, 4-B. They are characterised as "hormone" weedkillers as their selective herbicide action arises from their ability to change the growth rates of the plant weeds. In general herbicides in normal use present little toxic hazard to man but they may be highly toxic to aquatic life. An exception to this generalisation may, however, be 2, 4, 5, -T owing to the presence of a manufacturing impurity, dioxin, in preparations of this chemical. This chemical, suspected of being a teratogen, is present in variable amounts depending on the quality of the raw materials and the production process used.

The bipyridyl group of herbicides includes two important compounds, diquat and paraquat, which are widely used for ground clearance prior to sowing, thus doing away with the need for mechanical ploughing. They were first introduced about 1957 to 1958 and have subsequently found extensive application in agriculture. The chemical formulae of the two compounds are shown below:

I DIQUAT

II PARAQUAT

It will be noted that they contain two six-membered rings but with one of the carbon atoms of the benzene ring replaced by an atom of nitrogen forming pyridine. The two rings are linked directly without a bridging atom but are also joined through a bridge between the two nitrogens and involving two further carbon atoms. They are distinguished from most other herbicides by their non-selective and very rapid action on most plants. They are also quite rapidly inactivated on contacting the soil, so that the sowing of a crop can follow quite rapidly without the risk of damage to the crop or of its contamination by toxic residues.

The third class of herbicide compounds *the triazines* are based on another six-membered ring, but with three nitrogen atoms in this case replacing alternately three atoms of carbon. Some of these compounds were developed in Switzerland in the mid-1950s and they include compounds such as atrazine and simazine. They have the valuable quality of being selective weed-killers, especially for broad-leaved grasses, but without action on a maize crop; maize is resistant to triazines which it is able to metabolise rapidly and convert into harmless compounds.

The fungicide chemicals include several series of compounds, one of the classes including a number of both inorganic and organic compounds of mercury. A second group of compounds includes several metallic salts of the synthetic organic acid *dithiocarbamic acid*. Carbamic acid I is intermediate between the familiar carbonic acid and urea, and two atoms of oxygen are replaced by sulphur in dithiocarbamic acid II.

$$\begin{array}{ccc}
\text{HO}\diagdown & & \\
& \hspace{-1em}\text{C=O} & \\
\text{H}_2\text{N}\diagup & &
\end{array}
\qquad
\begin{array}{ccc}
\text{HS}\diagdown & & \\
& \hspace{-1em}\text{C=S} & \\
\text{H}_2\text{N}\diagup & &
\end{array}$$

$$\begin{array}{l}
\text{HS}\diagdown \\
\hspace{3em}\text{C=} \\
\text{H}_2\text{C}-\text{NH}\diagup \\
\hspace{0.5em}| \\
\text{H}_2\text{C}-\text{NH}\diagdown \\
\hspace{3em}\text{C=} \\
\hspace{2em}\text{HS}\diagup
\end{array}$$

I   II   III   HS

I CARBAMIC ACID   II DITHIOCARBAMIC ACID
III ETHYLENE BISDITHIOCARBAMIC ACID

   Two molecules of this acid linked through ethylene form the compound ethylene bisdithiocarbamic acid (III) from which most of the important fungicides in this group are derived. Their fungicidal action depends on their selective conversion by fungal spores into the corresponding highly toxic isothiocyanates.

   The fungicidal class of chemicals also includes a number of dinitrocompounds such as dinocap, binapacryl, and dinitro-cresol, and also two miscellaneous compounds known as captan and folpet.

   The compounds referred to above are representative of the principal classes of compound in general use and are by no means exhaustive. For a full listing and description of all possible compounds which have found applications in agriculture or in horticulture, reference should be made to the two publications sponsored by the British Crop Protection Council (3, 4) and the Agrochemicals Approved List, 1972, published by the MAFF.

PROBLEMS OF DETECTION AND MEASUREMENT

THE PROBLEMS presented to the analytical chemist in detecting and measuring pesticide residues can be imagined from the number, complexity and variety of the chemicals involved. The problem is complicated by the requirement to measure extremely small quantities at the most of a few parts per million. The problem is also aggravated by the contamination of a great variety of foods resulting from the transport and cycling of residues in the biosphere and the necessity to examine food imports when information about pesticide treatments abroad may be difficult to obtain.

The problems for the chemist are therefore essentially twofold:

1. The identification of any residues present
2. The quantitative determination of the actual amounts of any residues present.

The general pattern of analysis for a variety of samples of plant or animal origin involves three stages:

1. Separation of the residues from the sample – this generally involves maceration of the sample and its extraction into an organic solvent.
2. Cleaning up of the extract by means of column chromatography using alumina or by further partitioning between solvents.
3. The final qualitative or quantitative analysis of residues in the extract from 2.

For the final stage the old established paper chromatography technique or the more modern thin layer chromatography are most useful for chlorinated hydrocarbons, organophosphorus and chlorophenoxy compounds. The more recent application of gas liquid chromatography has provided the most versatile, sensitive and selective procedures. Infrared spectroscopy is invaluable for establishing the identity of a compound but the method suffers from a lack of sensitivity.

Because of the complex procedures involved in analysis complete and accurate determinations of significant residues are generally laborious and time-consuming. There have also been problems in the past in achieving a satisfactory resolution between chlorinated hydrocarbons and PCB residues.

CHLORINATED HYDROCARBONS IN THE ENVIRONMENT

WITH THE exception of preparations involving natural substances such as derris, pyrethrum, heavy metals and mineral oils the great majority of the organic chemicals developed as pesticides are artificial and do not occur in the natural environment. Their applications in agriculture and in public health to control certain disease-carrying insects and molluscs have introduced entirely novel factors into the natural environment. Although a great deal has subsequently been learnt about the behaviour of many of these substances in the environment there are still great gaps in knowledge and the full consequences especially those relating to any long-term effects cannot yet be defined very clearly.

DDT as the first of the chlorinated hydrocarbons was first produced during the Second World War essentially for health control measures in the forces. The scale of its production and subsequently of the other chlorinated hydrocarbons increased rapidly after the war, and their applications to agriculture and health control also multiplied rapidly. The total production probably reached a peak around 1964 when the world usage of DDT alone was reported to be about 400,000 tons with about one-fifth of the production used in malaria control. The production of both aldrin and dieldrin was also on a similar massive scale at about this time. Although the production of all chlorinated hydrocarbons has subsequently declined DDT especially has continued to dominate the problems of pesticides on the global scale. It has been estimated that the total amount of DDT manufactured to date is about 20 million tons. The problems of other pesticide chemicals are of a more localised nature, and with the possible exception of some of the formulations involving the heavy metals the effects are of shorter duration. Studies of the long-term distribution of the chlorinated hydrocarbons in the biosphere, their transport through food chains and the resultant concentrations in food and in man have inevitably received the greatest attention. Some of these studies will be reviewed in the remainder of this chapter with reference to some of the other chemicals where data is available. A consideration of the heavy metals is however deferred to Chapter 5.

Pesticide residues on a globul scale involve mainly the chlorinated hydrocarbons; apart from the scale of their application, food and human pullution problems are aggravated by their persistence in the environment and their tendency to become airborne and globally distributed. The order of their persistence in the environment is generally considered to be:

$$DDT = Dieldrin > BHC > Chlordane > Heptachlor > Aldrin$$

Aldrin is last in the series as it undergoes fairly rapid conversion into dieldrin. The first three in the series have been the most thoroughly investigated, and DDT in particular is the one type of residue which is found almost universally throughout the environment and in the principal food sources of man, usually in the largest amounts. (The term DDT as used subsequently will be taken to include its immediate degradation product DDE and other similar compounds which may be present as impurities in the original compound.)

The transport of DDT through the various parts of the environment ending in its presence in a food product is presented diagrammatically in Fig. 2.3. The concentrations of DDT at various stages of the food chains shown in the figure are typical values based in the main on results for the UK in 1968 and reported more systematically later in this chapter. They cannot be regarded as representative of all the situations which will arise in the natural environment but provide a general guide to the concentrations and the changes occurring at the various stages of a food chain.

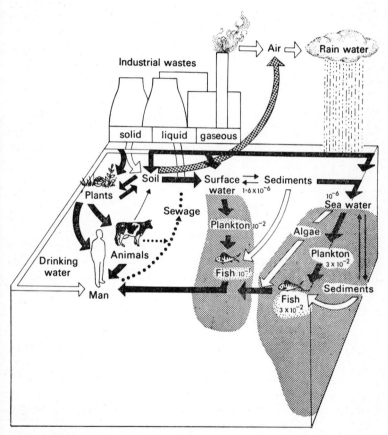

Fig. 2.3 DDT in food chains (DDT concentrations in ppm). The black arrows represent the critical pathways for food contamination

The ubiquity of DDT in the global environment today is due to the factors mentioned previously and its distribution is assisted undoubtedly by its tendency to become airborne. This latter tendency is due partly to its volatility and partly to its ability to adsorb on to the surfaces of

les of dust which can then become airborne and may remain
d in the atmosphere until washed out by falling rain. The
rations in rainwater falling in the UK are none the less extremely
veraging about 0·1 ppb; some of the DDT in the rain will be
applications of the insecticide in the UK and some will be due to
treatments in other countries. This type of deposition from rainwater, a
type of chemical fallout which is now occurring all over the globe will
always be minute compared with the deposition resulting directly from
the treatment of crops in any one year. An average rainwater concentration of 0·1 ppb would result in an annual deposition of just a few mg/acre
in Britain, which may be compared with a typical application rate of
about 2 lb/acre (about 2·5 kg/ha), which could lead to soil concentrations down to a depth of 6in. of about 1ppm – soil concentrations as
high as 0·3 ppm have in fact been reported in arable areas.

The deposition from rain falling into surface waters and into the
oceans will also be minute compared with the run-off from crops and
soils after treatment. The more direct routes of contamination are then
as shown below:

$$DDT \longrightarrow \begin{matrix} \text{soils,} \\ \longrightarrow \\ \text{plants} \end{matrix} \begin{matrix} \text{surface} \\ \longrightarrow \\ \text{waters} \end{matrix} \longrightarrow \text{oceans}$$

The pollution of the main sources of food in any one year will depend
therefore very largely on the scale of application to food crops in that
year. Although some accumulation may take place in the soil the transfer
of DDT from soil to plants via the roots is extremely small. In surface
waters such as large lakes and in the oceans, however, the accumulation
of DDT over many years may be more significant than any annual
increments. The accumulation of DDT in surface and sea waters from
fallout and from run off from soils and transport through rivers will
also be supplemented to some extent by industrial waste discharges and
careless disposal of agricultural containers.

The pollution in the fresh water and sea water ecosystems leading to
the pollution of fish, edible molluscs and seaweeds appears to have
been a greater hazard to fish life and to sea birds feeding mainly on the
fish than it has been to man, since fish and other edible products of the
sea contribute only a small part of the average total human diet. The
contamination of drinking water is also insignificant. The hazards to
fish and to birds are also quite serious owing to the large overall concentration factors from the water to fish and the bird population (*see*
Chapter 1, p. 13). Using the data from Fig. 2.3 the average concentration (accumulation) factor (C.F) for freshwater fish is seen to be about
66,000.

$$C.F = \frac{\text{concentration in fish}}{\text{concentration in water}}$$

The value for sea fish is about 30,000. The concentration in the livers of fish may reach about ten times the values in the whole fish, and even higher concentrations have been found in the livers of birds such as herons and grebes feeding on freshwater fish. These concentrations are high enough to harm the fish and bird populations and hence to damage the natural ecosystem with adverse consequences for human enjoyment and recreation. The concentrations in fish are also sufficiently high to cause some concern for small groups of the human population for whom fish may be an important part of the diet.

The principal dietary sources of pesticide residues are, however, agricultural products which provide about 90% of the average protein consumption in European countries. The total land under cultivation in England for example in 1966 amounted to just over 27 Macres, of which about one-half was pasture receiving practically no pesticide applications. Of the remaining land about 9·2 Macres was given over to farm crops, mainly cereal grains, about 0·35 Macres to vegetable crops, about 0·19 Macres to fruits and hops, and about 0·84 Macres to afforestation. The principal insecticides applied to the land were

1. γ-BHC, to about 3·7 Macres of farm crops mainly as seed dressings and dips
2. aldrin/dieldrin, about 0·54 Macres of farm crops again mainly as seed dressings and dips
3. DDT, about 0·19 Macres mainly fruit and vegetable crops.

Much smaller quantities of several other insecticide chemicals were also applied mainly to small acreages; they included endrin, endosulfan, derris, lead arsenate and lime sulphur.[5] The figures for 1966 also show an increase in the application of organophosphates compared with earlier years involving about 0·87 Macres of farm, vegetable and fruit crops and replacing the much more persistent chlorinated hydrocarbons.

The estimates of the areas of cultivated land affected directly by insecticide treatments provide no indication of the actual amounts of the different chemicals applied. The residual contamination levels in food will depend on the scale of application in addition to the acreage under treatment. The quantities of chemical required for field treatments by spraying and for seed dressings and dips are very different. Field treatments may involve the application of about 2 lb/acre (2·25 kg/ha) of the chemical, whereas seed dressings and dips would amount to no more than one-tenth of an ounce per acre (about 8 g/ha) or about

6000 lb per Macres. The comparative amounts of the three chlorinated hydrocarbons responsible for most of the treatments are therefore as follows: .

DDT, field treatment of   0·19 Macres $3·8 \times 10^5$ lb
BHC, dressings and dips 3·7   Macres $1·8 \times 10^4$ lb
Aldrin/dieldrin   ,,   ,,   0·54 Macres $2·6 \times 10^3$ lb

DDT is used almost exclusively in field treatments, the intensity of the application as measured by the quantity applied per acre, being greatest in orchards and vegetable plots. Concentrations as high as 10 ppm have been found in some orchard soils and about 0·3 ppm in arable land.

The presence of DDT etc in the soil may affect the natural micro-organisms which are involved in the beneficial processes of nitrification and conversion of organic matter but all the evidence suggests that present concentrations will have a negligible affect on these processes within the soil. There is also no evidence to suggest that the normal activity of earthworms is at all affected by the present concentrations of these insecticides. Earthworms consume large quantities of soil and organic matter, playing a vital role in the conditioning of the soil, and some field studies have shown that certain species are able to concentrate DDT by a factor of between two and three. This DDT may then be passed on to birds which feed on worms and other soil insects, and concentrations of several hundred ppm have been found for some orchard birds, such as blackbirds and thrushes. In these cases some of the DDT in the birds will also have come from feeding on the fruit. Some of the earliest evidence for the harmful effects of insecticide residues was of course provided by the large numbers of deaths among the seed-eating birds near agricultural land sown in the spring with insecticide dressed seed. Birds of prey such as the sparrowhawk and the peregrine falcon feeding on other birds and small animals were also badly affected by high concentrations of chlorinated hydrocarbon residues in their tissues. All of these effects may be detrimental to man's enjoyment of his natural environment and also serve as a possible warning to man himself.

It is to be expected then that the major part of the residues found in human food will arise directly from the treatment of crops, and mainly from the spraying of the growing crop. All residues, having a degree of persistence will tend to accumulate in the soil after being washed off the plants by rainwater or by the return of organic matter to the soil. The soil will therefore tend to act as an accumulator of the persistent chemicals. Some of the soil reservoir will be reduced by run-off as described earlier and some may serve as an additional source for growing plants through the roots. There have been very few studies attempt-

ing to relate the residue levels in plants to the soil levels, t
studies which have been carried out suggest that these resi
largely unavailable to the plant. The chlorinated hydrocarl
strongly bound to certain soil components, mainly clay mine
organic matter. The concentration factors giving the ratio of the plant
to soil concentrations are generally less than one. The plant is therefore
able to discriminate effectively against the pesticide chemical in the soil
and a factor as low as o·1 has been reported for carrots and potatoes
growing in a sandy loam soil. The transport processes involving the
residues in the plant have also received relatively little attention,
although it seems fairly clear that they tend to be mainly superficial;
in the case of root crops they can be reduced significantly by peeling
and by washing. There are some exceptions to plant discrimination
against pesticides in the soil, and these are most likely in the case of
crops such as rice where the roots are immersed in water; the situation
is then partially analogous to that prevailing in the freshwater
environment.

Most of the pesticide residues found on food crops arise therefore
from their direct treatment with the pesticide. The levels which are
found in the harvested crop will then depend to a large extent on the
time of application in relation to the maturity of the plant and its
harvesting, the climate prevailing after the application, and also on the
nature of the crop. The residues are again mainly superficial, being held
for instance in the waxy protective layers of fruits. In view of all these
factors the determination of average pesticide levels in crops for human
consumption is a matter fraught with considerable difficulties, as wide
variations are to be expected in the results for different regions and at
different times. A similar situation is expected to prevail with the results
for animal produce, with the fats of milk and dairy produce being
especially susceptible as accumulators of chlorinated hydrocarbon residues in the animal feedingstuffs. Most of the residue levels in animal
produce will actually result from the consumption by animals of contaminated feed supplements, as the average levels in pasturage are
normally quite low.

Very few investigations into the transfer of pesticide residues from
animal feedingstuffs into the tissues of farm animals appear to have
been carried out. A few trials carried out in the USA suggest concentration factors for milk over feed of less than one; about o·4 for dieldrin,
o·05 for DDT and o·14 for γ-BHC.[6] There is therefore a significant
discrimination against any pesticide residues entering into the animal
milk; most of the residues will be associated with the fatty content of the
milk, amounting to about 4% by weight. The concentrations which
are found in butter are about ten times greater than those which are
found in the fresh milk. This would suggest concentration factors for

Table 2.1

Residues of chlorinated hydrocarbon in principal foods, 1963–8[5, 7]

Food group (% of national diet)	Composition of group	Year	Residue levels — Total DDT Compounds	Dieldrin	ppm γ-BHC
Cereals (15·9)	63% bread, 12% biscuits, 10% cakes, 8·4% flour etc.	1966–7	0·018	0·003	0·014* ⎫ 0·009 ⎬
Flour		1968	0·009	0·001	0·018*
Breakfast cereals		1968	0·002	<0·001	0·004*
Fruits and preserves (16·8)	23% sugar, 21% soft drinks, 11·5% apples, 10% citrus fruits etc.	1966–7	0·026	0·002	0·005* ⎫ 0·002 ⎬
Root vegetables (16·4)	90% potatoes	1966–7	0·006	0·002	0·004* ⎫ 0·002 ⎬
Potatoes		1963–4	0·03	0·027	0·01*
Carrots		1963–4	—	0·037	0·05*
Green vegetables (8·2)	22% cabbage, 22% canned 15% tomatoes, 6% salads	1966–7	0·013	0·003	0·06* ⎫ 0·03 ⎬
Meats (14·5)	16% eggs, 16% mutton and lamb, 14% beef and veal 14% pig meats, 10% fish	1966–7	0·050	0·009	0·017* ⎫ 0·007 ⎬
Beef kidney fat (UK)		1966	0·07	0·04	0·05*
Mutton kidney fat (UK)		1966	0·11	0·44	0·16*
Corned beef		1964	<0·01	0·01	0·03*
Poultry (UK)		1965, 66	0·02	<0·01	0·06*
Eggs (UK)		1965, 66	0·1	0·01	0·03*

Fish (UK)	67% battery, 25% deep litter, 8% free range	1968	0·032	0·007	0·004*
Fats (4·9)	26% butter, 17% cheese		0·21	0·024	} 0·059**
	14% ice cream, 12% margarine, 10% lard	1966–7			0·013
Butter (UK)		1968	0·06	0·03	0·08*
Cheese		1968	0·03	0·01	0·05**
Lard		1968	0·17	0·02	0·03*
Cooking oils		1968	0·04	<0·01	0·01*
Cod liver oil (Arctic)		1968	1·65	0·16	0·09*
Milk (23·3)	100% liquid milk	1966–7	0·004	0·002	} 0·003*
					0·001
Whole milk (UK)		1966	0·004	0·002	0·006*
Dried full cream milk (UK)		1966	0·028	0·011	0·04*

* Total BHC compounds

57

butter of about 4 for dieldrin, and 0·5 for both DDT and γ -BHC. The highest pesticide residues in any foodstuffs occur in animal fats, and especially lard, but values in the processed fats such as margarine are quite low. The processing of oils into margarine evidently leads to significant reductions in any pesticide residue levels.

CHLORINATED HYDROCARBON RESIDUES IN FOODS

THE RESULTS of several UK surveys of residue levels of DDT, together with dieldrin and BHC compounds in major foodstuffs are presented in Table 2.1. (5, 7)

Table 2.1 summarises results for a variety of individual staple foods and food groups measured by official laboratories in the UK at the request of the Advisory Committee on Pesticides and other Toxic Chemicals. The results have been published in two series of papers in the Journal of the Science of Food and Agriculture, starting in December 1966,[7] and are summarised in the Advisory Committee's Further Review of Certain Persistent Organochlorine Pesticides used in Great Britain.[5]

The results for the main food groups of the composition shown in column 2 are based on a total diet study during the years 1966–7. This study was designed to determine the average pesticide levels in the major food groups of the diet after their preparation and cooking in readiness for consumption. For the purposes of this survey the various foods were divided into the seven groups shown in the table (see also p. 17).

The other results are for individual staple foods within the groups and selected by the Panel of the Advisory Committee for their importance in the individual dietary; they form part of a continuing survey of residue levels in the UK.

Some analyses of the fruit from blackcurrant bushes treated with endosulfan and representing about 80% of the crop show residue levels of 0·31 ppm in 1966 and 1·17 ppm in 1967, but the values are reduced approximately one hundredfold in blackcurrant preserves. The results for arsenic, lead and mercury residue levels also based on the surveys of individual foods are presented in Chapter 5.

HUMAN INTAKES OF CHLORINATED HYDROCARBONS

THE RESULTS of the 1966–7 total dietary survey in the UK provide the best estimates of the average total daily consumption of the principal chlorinated hydrocarbons; this amounted to 67·5 μg/day with DDT compounds accounting for just two-thirds of the total, Table 2.2.

Table 2.2.

Average daily consumptions of chlorinated hydrocarbons, 1966–7[7]

Food group	% by weight of total food consumed ca 1·7 kg/day	DDT Compounds	μg Dieldrin	BHC Compounds
Fats	4·9	17·1	1·9	4·9
Meats	14·5	11·8	2·2	3·9
Fruits, preserves	16·8	6·7	0·4	1·4
Cereals	15·9	4·4	0·6	3·5
Root vegs.	16·4	1·6	0·5	1·0
Green vegs.	8·2	1·5	0·3	0·7
Milk	23·3	1·3	0·7	1·1
Totals	100	44·4	6·6	16·5

Estimates of the dietary consumption in other years can only be inferred from results for selected foodstuffs in those years. These results suggest that the consumption was greater in 1964 and declined through 1966/1967 to 1968. The DDT consumed in 1968 was probably in the range 25 to 30 μg/person/day compared with the figure of 44·4 μg/person/day in 1966/1967. All values are average estimates and individual consumptions may vary significantly depending on the composition of the foods consumed and their sources. Dietary surveys have also been carried out in the USA in recent years and suggest that the average daily consumptions per person in that country may have been about three to four times higher than in the UK, mainly owing to higher residue levels in animal produce.

The high contributions of fats and meats to the total dietary consumption of the three insecticides should be specially noted. These two food sources are together responsible for just over 60% of the DDT and dieldrin consumed in each case and about 55% of the BHC; the importance of fats despite their contributing less than 5% of the daily dietary, is especially significant. The relatively high concentrations of these pesticides in meat is most probably due to their presence in the fatty parts of the meat; the chlorinated hydrocarbons are distinguished by their solubility in fats, and it is to be expected that this association will be maintained even after human consumption, metabolism, and deposition in the body.

The residues of the chlorinated hydrocarbons entering the body in food are absorbed relatively slowly in passing through the small intestine and are then transported rapidly to the liver via the portal vein. Within the liver they undergo biotransformations quite slowly, mainly owing to the stability of the bond between chlorine atoms and the carbon atoms in the ring structure. They are, therefore, totally foreign substances to the liver and the enzymes for breaking down those bonds do not normally exist. They are, however, slowly transformed into a number of metabolites mainly by processes involving more vulnerable parts of the molecular structure (Fig. 2.4). The diagram illustrates the transformations of DDT compounds including DDE into the more polar com-

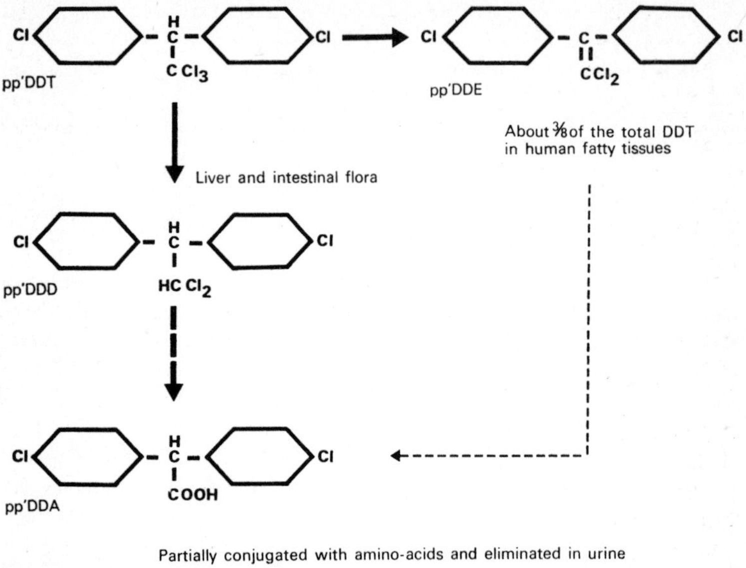

Fig. 2.4 Metabolites of DDT in man

pound DDA by the action of the liver enzymes and to some extent by the intestinal flora. The acidic DDA can combine with certain amino-acids and is then more readily transported and eliminated. Small quantities of the metabolites may also be secreted via the bile into the small intestine, where they may be recycled through the small intestine or eliminated in the faeces. Aldrin is rapidly transformed into its epoxide dieldrin, but the latter is only slowly transformed into various metabolites by oxygenation or hydroxylation.

The original compound or its metabolites may remain in the liver or enter the bloodstream and small quantities are then filtered out and eliminated in the kidneys. Any substances remaining in the bloodstream are initially distributed among all the body tissues followed by a slow re-distribution and concentration in the adipose tissues. Once absorbed into fat they are relatively immobile and harmless until such time as the body's reserves may be called upon at times of stress or in illness. The levels in fat increase over a period of time, when there is a steady intake, and eventually reach an equilibrium value in the case of DDT after about eighteen months. The equilibrium amount of DDT in body fat is proportional to the average daily intake and at this stage the elimination rate has increased to balance the rate of intake; if the intake falls to zero the elimination rate gradually drops with a halving time for the actual concentration in fatty tissue of about one year. Concentration factors in fat and in human milk at the steady state are summarised in Table 2.3.

Table 2.3
Concentrations of chlorinated hydrocarbons in human milk and fat 1965–7

	DDT Compounds	Dieldrin	BHC Compounds
Human fat μg/kg *	2900	200	300
Amount consumed μg/day	44·4	6·6	16·5
Average concentration in diet, μg/kg (1·65 kg food/day)	27	4·1	10
C.F $= \dfrac{\text{conc. in fat}}{\text{conc. in diet}}$	110	49	30
Human milk μg/kg†	130	13	6
C.F $= \dfrac{\text{conc. in milk}}{\text{conc. in diet}}$	4·8	3·2	0·6

* Report of Advisory Committee[5]
† Values for 1963/64 quoted by C. A. Edwards[8] (dietary concs. in these years assumed to be similar to 1965/67)
N.B. About two-thirds of the total DDT is in the form of DDE

They show high concentration factors in the fatty tissue from the dietary intake. The concentration factors are of course much less if the human concentrations of the residues are expressed over the total body weight but this is to give a misleading indication of the ability of the body to concentrate the residues in the fat. Human fatty tissues, however, may not be the critical tissues of the body in the sense that they are at the greatest risk from any harmful effects which may be due to these higher concentrations of the residues. DDT and the other chlorinated hydrocarbons in the fatty tissues are removed for the time being from harm's way, but may be released at times of stress and especially reduced intake of food and complicate the condition of the patient. The situation is very different from the accumulation of strontium-90 in bone (Chapter 4); the bone not only has the highest concentrations of strontium-90 in body tissues but the bone marrow is most at risk from the radiations emitted by this fission product. The fatty tissues do, however, serve as the body's reservoir of DDT and the other compounds and in this sense the concentration factor from diet to fat cannot be ignored. The high values for DDT are to be specially noted as also is the concentration factor for human milk which may have significance for breast-fed children.

The average concentrations of the chlorinated hydrocarbons in various tissues of the body have been found to have the following ratio:

fat/liver/kidney, brain, gonads, blood $= 100/10/1$

The actual storage levels in the human tissues increased steadily from the time of the first large-scale applications in the post-war period, probably reaching maximum values in the early 1960s. Voluntary restrictions and government bans on the application of these chemicals should now be resulting in a steady decline in the concentrations.

The data which are summarised in Tables 2.2 and 2.3 provide a useful guide to average levels in food and human adipose tissue in the UK in the latter part of the 1960s. Examination of similar data for the USA shows that significant differences in the levels of DDT and its metabolites in human fatty tissues exist and can be attributed to differences in individuals and also to the location or place of exposure. The sex, race, age and socio-economic circumstances of an individual all appear to have some bearing on the DDT levels found in the USA, whereas there is a much greater uniformity in the levels found for dieldrin. Average levels for dieldrin in fat show much less individual variation and therefore have greater significance as national levels. The variability in DDT levels has raised some doubts as to whether dietary sources are the predominant source of this pesticide or whether other sources such as contaminated dust in the home from domestic applications may also be significant but of variable importance. The values reported for average dietary consumption can therefore only be regarded as a guide to the recent levels of human exposure with appreciable variations in individual exposures.

The average concentration factors which have been estimated from the surveys in the UK, Table 2.3, are found to be comparable with the factors which can be estimated from similar total diet studies in the USA. A further comparison with these values is afforded by the results of trials in the USA, involving a number of human volunteers. The volunteers were fed an average of 35 mg of technical DDT each day, equivalent to a concentration in their diet of about 21 ppm. During a period of 21·5 months the concentrations in the fatty tissues were found to reach a steady value of 281 ppm, giving concentration factors from diet to fat of 13·5. When purified DDT was substituted for the technical material the concentration factor increased slightly to 16. Other volunteers consumed an average quantity of 0·211 mg dieldrin per day, in addition to 0·014 mg already present in the food, giving a total dietary concentration of 0·13 ppm. During a period of two years in this case the concentration in the fatty tissues reached a steady value of 2·85 ppm, equivalent to a C.F (fat/diet) of 21. Workers at a DDT manufacturing plant, estimated to be consuming about 18 mg DDT/day were found to have fat concentrations from 38–647 ppm, equivalent to a C.F (fat/diet) between 3·5 and 59.[9]

The results for the DDT concentration factors (fat/diet) based on the trials with human volunteers and on persons exposed to DDT regularly in the course of their work are all lower than the values which have been calculated for the exposures of the general public, and based on the national surveys. It would appear that there is more discrimination by the body against DDT when it is added deliberately to a diet or is ingested at work. In the latter case absorption through the lungs or through the skin could also be a significant difference. The entry of DDT into foods as a result of agricultural operations could also lead to a much more intimate incorporation of the DDT into the food than would the case with the volunteer feeding trials. Alternatively the difference may be associated with chemical factors and the presence of variable amounts of DDE in the foods.

THE POLYCHLORBIPHENYLS

THE POLYCHLORBIPHENYLS (PCBs) have a much longer history of manufacture and usage than the chlorinated hydrocarbon pesticide chemicals. Their toxicity was first noted as early as 1919 and it has been estimated that by 1939 six human deaths had occurred as a result of industrial operations with these materials. Their importance within the environment and their persistence only became apparent, however, as a result of the investigations into the chlorinated hydrocarbons with which they are often closely associated and sometimes mistaken. The major environmental problem from PCBs arises from industrial effluents discharged into rivers or the sea. Like the chlorinated hydrocarbon compounds they enter readily into various food chains in the fresh or marine water environments with accumulation taking place at each stage of the food chain. Quite high concentrations have been found in the top species of a food chain with values as high as 900 ppm measured, for instance, in herons in the UK. In the human diet the highest concentrations are found in freshwater and marine fish with lower levels in milk. An intensive survey of PCBs in food was carried out in Sweden between 1967 and 1969 and some of the results have been quoted by R. Edwards.[10] All milk samples were found to have concentrations less than 100 ppb. The average value in butter was 30 ppb, compared with a concentration of DDT in butter samples in the UK survey of 60 ppb (Table 2.1). The concentration of PCBs in the yolk of eggs was similar to the butter concentration. The average concentrations in the total food consumed and in the human body are probably not very different from the levels reported for DDT. Average concentrations in human milk found in the Swedish survey were found to be 160 ppb, which is only just slightly higher than the DDT concentration in human milk reported in the UK survey, Table 2.2. The metabolism of PCBs in the body is probably very similar to that of DDT and their biological effects at low concentrations are also expected to be similar. The greater variability and uncertainty in their composition and especially the extent of their chlorination creates problems in assessing their toxicity to various organisms and to man. The possibility of the presence of small amounts of more highly toxic components in a commercial mixture of PCBs cannot therefore be entirely ruled out.

OTHER COMPOUNDS IN THE DIET

(a) Organophosphates

THE RESIDUES of chlorinated hydrocarbons in diet have been the subject of most of the surveys in view of their undoubted persistence at all levels of the environment. There have been considerably fewer analyses for other types of residue including the potentially highly toxic organophosphates which are finding increasing application as replacements for DDT. Their concentrations in foodstuffs may be expected to show more variation owing to their more rapid decomposition in the environment and the more sporadic nature of their application. Determinations of their average dietary intakes by man will be correspondingly less reliable.

The results of some analyses on foodstuffs from the report of the Joint Survey of Pesticide Residues in Foodstuffs are reproduced in Table 2.4. They show that residues of organophosphates are fairly widespread; their presence in samples of prepared foods such as bread and sausages also suggests that they may be rather more persistent than is generally assumed to be the case.

Table 2.4

Organophosphorus residues in selected foods ([11a])

Food	Organophosphorus residues (as P) (ppb)
Bread	140
Imported cheese, soft	700
Pork	1370
Milk	80
Infant food (milk based)	125
Sausages	60
Breakfast sausage or salami	60–180
Ham	60
Lettuce	87

Malathion appears to be the particular organophosphate most frequently encountered. It has, for instance, been shown to be present in samples of imported grains at concentrations ranging from 3 to 5 ppm[11] . Much of the malathion appears to be associated with the dusty material which is partly separated from the grain during its transfer from ship to silo or at the flour milling plant. Some of the dust inevitably remains attached to the grain and finds its way into animal feedingstuffs in sufficient amounts to account for its presence in animal food products. Other organophosphates, used as insecticides in treating fruits and vegetables, were only very occasionally found in the produce at levels in excess of the limits of detection of the analytical procedure employed. These levels should always be very low if the instructions for ensuring a safe period of time between application and harvesting of the crops are correctly observed.

The total dietary survey referred to in the section for residues of chlorinated hydrocarbons was extended to include the organophosphorus compounds.[11a] The survey established that only a very small number of the samples from the seven food groups contained residue levels in excess of 0·02 ppm. Combining the results of the Joint Survey, Table 2.4, with those of the total diet study indicates a maximum possible concentration of 0·1 ppm organophosphates in the total diet. This is almost certainly an overestimate and actual concentrations are likely to be considerably less. The average dietary intake in the USA was for example estimated to be about $11\mu g$ per day, of which 80% was identified as malathion. There are no grounds to suspect higher intakes in Britain; a dietary intake of $11\mu g$ per day would be equivalent to a concentration in the food consumed of no more than 0·01 ppm. The human contamination levels are almost certainly extremely low.

Organophosphorus compounds are quite readily absorbed through the small intestine and transported via the portal vein to the liver. Any compounds reaching the liver are converted rapidly into non-toxic mineral phosphates. The only possible exception is parathion where small amounts of a more toxic metabolite may be formed temporarily; its toxicity to insects depends on such a conversion. The final process of detoxication is rather more slowly accomplished, but there is no evidence that residues of this compound occur in food samples in the UK.

(b) The auxin herbicides

THESE COMPOUNDS are widely used as herbicides with 2, 4-D being the most important of the group on a global scale. Formulations of these compounds were also used on a massive scale by the American forces in Vietnam from 1962 to defoliate the jungle and to deny crops to the enemy. The most important of these compounds from the stand-point of human health is undoubtedly 2, 4, 5-T, largely owing to the probable presence of the impurity, dioxin, produced in the manufacturing process. This compound is known to be a powerful teratogen and to be highly toxic to laboratory animals. Its LD50 for guinea-pigs is stated to be only 0·0006 mg/kg body weight, which may be compared with the much higher values for 2, 4-D and 2, 4, 5-T (see below, Table 2.6).

The compounds are fairly widely used in the UK on about 70% of the acreage devoted to cereal grain cultivation, and on a scale of treatment which is comparable with the applications of the insecticides and fungicides. They are however simpler compounds than the organo-chlorine insecticides and they are more readily bio-degradable and present at much lower concentrations in food. But relatively little is known about their persistence in the environment or their transport through food chains but there is no evidence to suggest that they constitute a global pollution problem as in the case of DDT. A Herbicide Assessment Commission of the American Association for the Advancement of Science is at present investigating the consequences of the military applications in Vietnam, and is especially concerned with the concentrations of dioxin in the local diets. This work has however been handicapped so far by the lack of sufficiently sensitive analytical methods. There is very little data available to enable an assessment of the concentrations in a normal diet but Way,[12] has summarised the results of a survey carried out in the USA. The results of this total dietary survey conducted in 1964–1966 showed an average daily intake of about 10 μg/day with one-third of the intake being 2, 4-D and one-half MCPA, milk and preserves and to a lesser extent cereals being the major food sources. Very little also appears to be known about the fate of these substances in the body; they appear to be fairly rapidly absorbed and then largely eliminated via the kidneys and urine. Apart from the possible presence of dioxin their toxicity to man is expected to be low. Their toxicity to trout in a freshwater environment and at the agricultural rates of application appears to be about one-hundredth of that due to aldrin. More information about the effects of these compounds on man is expected when the full reports of the American studies in Vietnam are published.

The application of these compounds to crops may also have certain consequences for grazing animals. The nitrate content of the leaves of the sugar beet for example may be increased about tenfold after treatment with the herbicide, and this may be a factor in some cases of nitrate poisoning of stock. Other changes in the composition of plants may also occur such as an increased content of certain alkaloids or of prussic acid and these could be of some human significance.

BIOLOGICAL EFFECTS OF PESTICIDE RESIDUES

THE HARMFUL effects of any pesticide chemical involve a complex of factors. In addition to the magnitude of the exposure and the rate at which the exposure occurs, any effects will depend on the critical pathways of the foreign substances within the body, on any biotransformations occurring along the pathways, and especially on the target organ in which the original substance or any toxic metabolite is concentrated preferentially or which may be especially sensitive to damage. The state of health enjoyed by the individual and the adequacy of the diet may also be factors in modifying and adapting to any toxic effects.

All pesticide chemicals are selected for their biological activity and their ability to be toxic to and hence destructive of some form of pest, insect, fungus or weed. The various classes differ appreciably in their chemical behaviour and mode of action within the organism. These differences are illustrated in Table 2.5; some differences between individual members of a group may also occur, and it may not always be

Table 2.5

Pesticide modes of action

Mode of action	Chemical class
Inhibition of photosynthesis	Triazines Carbamates Bipyridyls
Inhibition of acetyl- cholinesterase	Organophosphates Carbamates
Hormone analogues	Chlorinated phenoxy-acetic acids
Neurotoxicants	Chlorinated hydrocarbons Pyrethrins

possible to infer the properties of a whole class from one or two specially investigated members of the whole class.

Among the insecticides, the chlorinated hydrocarbons are essentially neurotoxicants, damaging the nervous system of the organism; the organophosphates and carbamates are specifically inhibitors of the enzyme acetylcholinesterase which controls a vital link in the sequence of chemical processes involved in the transmission of nerve impulses and can therefore lead to paralysis of the nervous system. The herbicides, on the other hand, either interfere with the process of energy fixation (photosynthesis) or with processes dependent on energy utilisation in plants. The chlorophenoxy-acetic acids act as hormone analogues or growth promoters.

The metabolic processes which are carried out in the various tissue cells of all multi-cellular organisms have very many features in common and it is not surprising therefore that the same chemicals can also prove damaging to animals and to man. There have indeed been numerous incidents involving accidental high level exposures to a number of pesticide chemicals where acute symptoms of poisoning have occurred with fatalities in some cases. The more serious outbreaks of poisoning due to pesticide chemicals have generally involved organophosphates, such as parathion, or organo-mercury compounds. A few fatal accidents have also been reported with two of the chlorinated hydrocarbons, endrin and BHC. Accidents have occasionally involved young children when chemicals have been introduced into the home and stored in the wrong container, such as soft drink bottles. In some other cases, epidemics of poisoning have occurred when empty chemical containers have been wrongly used for the storage of food. Further tragic cases of poisoning have also occurred when cereal grains dressed with fungicide have been accidentally diverted into food supplies (*see* Chapter 5). Acute exposures of this nature serve to demonstrate the essentially toxic nature of these chemicals and some of the dangers inevitably involved in their application. The relative toxicities of some of these chemicals at the acute level of exposure are also illustrated by the values of the LD50, the lethal dose 50% (Table 2.6). All these values have

Table 2.6

Some examples of the LD50 (semi-lethal dose) for pesticide chemicals in rats[9]

Group	Chemical	LD/50, mg/kg body weight
Organochlorine	Dieldrin	46
	DDT	113
	BHC	88
Organophosphates	Dichlorvos	100
	Malathion	2800
	Parathion	6·4
Chlorphenoxy-acetic acids	2, 4-D	400
	2, 4, 5-T	300
Dithiocarbamates	Maneb	7500
Carbamates	Carbaryl	850

NOTE
The LD50 (lethal dose 50%) column 3 is that concentration of the chemical which when present in a large number of the animals results in an average of 50% fatalities

been obtained from feeding trials on experimental animals and cannot, therefore, be applied directly to human toxicity although it is reasonable to expect that they provide a relative guide to toxicities. Neither these values or the evidence of acute effects arising from accidental exposures provide any guidance at all regarding the long term chronic effects due to the small exposures at slow rates resulting from the consumption of contaminated food. In view of the differences in the biological activities of the various classes of chemicals they are likely to produce a variety of responses in man and other animals; differences in the human toxicology are to be expected therefore for the different groups and will also be very dependent on the amount of and the rate at which the exposures take place.

(a) Chlorinated hydrocarbons

THE CHLORINATED hydrocarbons have probably received more attention than any other class of chemicals because, in addition to their large-scale use, their persistence and their ability to concentrate in the fatty tissues of higher members of a food chain, some of them, especially DDT and dieldrin, are also present in the liver and adipose tissues of man at higher levels than the chemicals of any other pesticide group and contaminated food is the major source of their intake.

The effects of chlorinated hydrocarbons on the body depend very much on the quantity consumed. Severe acute poisoning and some deaths have occurred as a result of a very few accidental exposures involving endrin and BHC but DDT does not appear to have been involved in any fatal accidents. They are known to have an effect on the central nervous system leading to serious disorientation in the case of experimental animals. Some comparatively mild adverse reactions have been reported from a few human volunteers receiving a single acute dose of 750–1000 mg DDT. Maximum symptoms occurred some 6 hours after dosing and included a general feeling of malaise with some uncertainty of gait, a cool moist skin and increased sensitivity to contact. There were no detectable changes in the reflexes and recovery was complete after 24 hours. A higher consumption of 1500 mg DDT produced a slightly more severe reaction with giddiness, confusion and tremors of the limb extremities after 10 hours but complete recovery again after 24 hours. Single acute exposures, however, do not represent in any way the chronic situation affecting the general public from the small exposures received from food over a long period of time. A far more relevant series of exposures were arranged with a number of volunteers consuming either 35 mg/day of DDT or 0·225 mg/day of dieldrin for periods of from 18 to 21 months (see p. 62). Maximum fat concentrations of 400 ppm DDT and 3 ppm dieldrin were reached during this time. Regular examinations of the functioning of the liver and the central nervous system were carried out for the duration of the experiment and for two years after its conclusion without any adverse

effects being detected. The small group of workers on a DDT manufacturing plant, consuming 18 mg/day for 20 years also showed no adverse
clinical effects in that time.[9] These trials suggest that serious short-term
effects are unlikely especially at the present dietary consumption levels,
but the possibility of longer-term chronic effects still cannot be ruled
out on the basis of these trials and in view of the evidence with
experimental animals (Table 2.7).

The high toxicity to fish at quite low concentrations in water, and
the serious declines in the populations of some birds such as the peregrine falcon and sparrowhawk from 1955 to 1963 must also serve as a
warning to the human population. The deaths of large numbers of
birds feeding on aldrin/dieldrin treated grain seeds sown in the spring,
and also some orchard birds provide additional warning signs. It has
been suggested that concentrations of more than 10 ppm dieldrin or
30 ppm DDT in tissues other than fat may be fatal to several species of
bird. The presence of even lower concentrations in the eggs of certain
birds of prey appears to be associated with reduced thickness of egg
shells and an increased breakage rate by the parent birds. This could
be a factor in the decline of several species due to a reduction in breeding
successes. There seems to be fairly conclusive evidence therefore that
the chlorinated hydrocarbons have been hazardous to bird populations.
The deaths of foxes and badgers have also sometimes been associated
with high residue levels.

The environmental problems of the chlorinated hydrocarbons are
further complicated by the presence of the similar group of PCBs,
some of which could be even more toxic.

(b) General chronic effects

THE EVIDENCE of the feeding trials with DDT and dieldrin described
in the previous paragraphs provides some information about adverse
effects at sub-acute levels of exposure but provides no evidence whatever for any long-term chronic effects which might be expected from
exposures continuing at the present levels in food. The evidence for
long-term chronic effects is based entirely on animal experiments
involving levels of exposure which are still well above food consumption
levels and has been very thoroughly reviewed by the Mrak Commission.[9]
An attempt to summarise their essential findings for long-term carcinogenic, mutagenic, teratogenic and other effects through enzyme interactions is set out in Table 2.7. It has to be emphasised that these trials
with animals have been conducted at dosage levels well above the levels
received by man from his food; the data in all cases, therefore, are
inadequate to enable the incidence of any effect to be determined, even
in the experimental animals, at the levels which would correspond with
the human dietary intakes. Even if this could be ascertained on the

Table 2·7 Some chronic effects of pesticide chemicals

Group	Name	Mutagenicity Prob.	Poss.	Carcinogenicity Prob.	Poss.	Teratogenicity Prob.	Poss.	Enzyme interaction Prob.	Poss.
Chlorinated hydrocarbons	DDT	X		X				X	
	Dieldrin (Aldrin)	X		X		X		X	
	Heptachlor			X				X	
	Endosulfan				X			X	
	Endrin	X						X	
Organo-phosphates	Malathion		X						
	Parathion		X				X		
Carbamates	Carbaryl	X				X		X	
Dithiocarbamates	Maneb				X				
	Nabam				X				
	Thiram	X			X		X		
	Ziram	X			X				
Triazines	Simazine		X		X				
	Atrazine		X		X				
Phenoxy-acetic acids	2, 4-D	X			X		X		
	2, 4, 5-T	X			X	X			
Bipyridyls	Diquat		X						
	Paraquat		X						
Miscellaneous	Chlorfenson				X				
	Captan				X	X			
	Folpet				X	X			

NOTE The data summarised in the table are based on the report of the Mrak Commission[9]; the evidence for an effect is based overwhelmingly on animal experiments – "probable" signifies that this evidence is adequate to suggest a comparable human effect. "Possible" – the evidence is not totally conclusive and extrapolation to humans increasingly problematic basis of the animal trials, there would still be the further problem of interpreting the significance of the results for man. The evidence, therefore, is totally inadequate to establish the magnitude of any human long-term effects or whether there exists a threshold level below which it could be confidently stated that no effects will occur. The best that can be done, therefore, is to treat any evidence when it is soundly based as presumptive for similar effects in man. The Commission, when reviewing the data for three of the chlorinated hydrocarbons (DDT, aldrin and dieldrin) is satisfied that the results are sufficiently convincing for them to be regarded as "positive for tumor induction". This can be regarded only as presumptive evidence for an effect of unknown magnitude in man. It does, however, carry a strong assumption that human exposure to these chemicals must be minimised and restricted effectively to those applications where the benefits to man are unquestionable. The table indicates a "probable" effect for man when the data are judged to be acceptable and based on satisfactory testing procedures. In those cases where an effect is marked as "possible", the Commission regards the experimental evidence to date as insufficient and has made a strong recommendation for further trials as a matter of some urgency.

An additional attempt at an assessment of human effects from chronic exposures can be based on epidemiological studies in human populations. These studies attempt to identify an increased incidence of certain chronic effects in groups of the population who are known to have been exposed to significantly higher levels than the average values for the whole population. These higher levels may be due to occupational exposure or to the proximity of groups of people to areas where the chemicals have been applied on a large scale. The few studies along these lines have again been inconclusive and the Commission is forced to conclude that it cannot establish the incidence of any long-term chronic effects in man.

The problems of identifying those interactions involving enzyme activity and leading to either an increase or a decrease in the biological activity of a chemical agent by prior or simultaneous exposure to another agent are even more formidable. As these interactions may also involve chemicals used as food additives or medical products, the testing of all possible combinations is clearly out of the question. The essential need, therefore, is for a clearer understanding of the precise mechanisms which are involved in the various interactions. In the meantime it has, however, been established that the chlorinated hydrocarbons, and especially DDT are potent enzyme-inducing agents in the liver. The activity of these metabolising enzymes in the liver is, therefore, enhanced and this may have a pronounced effect on the activity and efficiency, for example, of certain drugs. There is also the possibility of the converse effect when a drug might be administered to accelerate the elimina-

Table 2.8 Recommended ADIs and practical residue limits for selected pesticide chemicals[13]

Pesticide group	Pesticide chemical	Max. ADI[a] mg/kg body wt.	Practical residue limits[b] Food	Limit ppm	Tolerances[c] Food	ppm
Chlorinated hydrocarbons	DDT	0·005*	Whole milk	0·05	Soft fruits, apples, peas, green veg., meat, poultry (fat basis)	7
					Strawberries, root veg, nuts	1
	Dieldrin (Aldrin)	0·0001	Eggs (shell free)	0·5	Potatoes	0·2
			Whole milk	0·15	Green veg., fruit (other than citrus)	0·1
			Eggs (shell free)	0·1	Citrus fruits	0·05
			Lettuce, carrot	0·2	Whole rice	0·02
	γ-BHC (Lindane)	0·0125	Whole milk	0·1	Cereals	0·5
			Eggs (yolk)	0·2	Veg., cherries, grapes, plums, strawberries	3
					Animal fat	2
Organo-phosphorus	Malathion	0·02			Flour, green beans, apples, green veg., other fruits	2 – 8
	Parathion	0·005			Fresh fruit and vegetables	0·5 – 1·0
	Dichlorvos	0·004			Milk, animal produce, fruit, cereals	0·1 – 5·0
Carbamate	Carbaryl	0·01			Fruits and vegetables	5·0 – 10
Dithiocarbamates	Maneb	0·025				
	Thiram	0·025				
Chlorphenoxy-acetic acids	2, 4-D	0·3			Cereals	0·2
	2, 4, 5-T	–				
Bipyridyls	Diquat	0·002			Edible oils, root veg, cereals	0·1
	Paraquat	0·0007			Potatoes	0·1

NOTES (a) Maximum ADI (acceptable daily intake) in mg/kg body weight: defined as the daily intake which during an entire lifetime appears to be without appreciable risk on the basis of evidence available at the present time.

(b) Practical residue limit: the maximum unintentional residue (i.e. arising from circumstances other than the protection of a food) allowed in a food.

(c) Tolerance: the maximum concentration of a residue permitted in a food as soon as practicable after harvest and prior to processing; for meat and poultry the DDT tolerance is given for the fat content. For specific values the original reference should be consulted.

* ADI "conditional" subject to further review

tion of certain pesticide chemicals. The drug phenobarbital increases liver enzyme activity; it might speed up the normally slow transformation rates of the pesticides, hastening their elimination and reducing the storage levels in the body. This procedure has, in fact, been recommended for the treatment of a dairy herd which had been accidentally exposed to high levels of DDT.

There is also the risk that any increase in enzyme activity may potentiate the effects of another agent by increasing the concentration of a more toxic metabolite. The Mrak Commission has again reviewed all the available evidence and has reached the conclusion that in the case of DDT the present intake levels are not sufficient to cause significant interactions due to enzyme induction. The Commission have also stated that although interactions between pairs of pesticide and drug chemicals, or between pairs of pesticide and food additive chemicals, are all possible, there are no grounds for genuine concern at the present moment. The need for further investigations and for better understanding of the basic mechanisms of interaction is, however, underlined both by the Commission and the World Health Organisation.[13]

ACCEPTABLE DAILY INTAKES OF PESTICIDE CHEMICALS

RECOMMENDATIONS FOR the Acceptable Daily Intakes (ADIs) for most of the pesticide chemicals referred to in the earlier pages of this chapter are summarised in Table 2.8. They are extracted from the 1971 Report of the Joint FAO/WHO Meeting on Pesticide Residues in Food.[13] The recommendations are based on the Meeting's assessment of all the relevant toxicological and related data. It will be noted that the ADIs are expressed as mg/kg body weight, so that in the case of the standard man weighing 70 kg, the maximum allowable intake of DDT would be 350 μg per day, and for dieldrin it would be 7 μg per day. These recommended values may be compared with the average UK assessments of daily consumption, giving about 45 μg per day of DDT and 6·6 μg per day of dieldrin, during 1966–7. The actual DDT intake is clearly well below the acceptable level but that of dieldrin is rather close to the ADI. As the food contamination levels of dieldrin have been declining since about 1964 owing to restrictions on its use in agriculture, it should now be well below what is considered to be an acceptable intake (Fig. 2.5).

The table also includes the Meeting's recommendations for practical residue limits and tolerances for selected foodstuffs. These concepts as defined by FAO/WHO are given at the foot of the table. ADIs should not be exceeded if the recommended practical residue limits and tolerances are not exceeded.

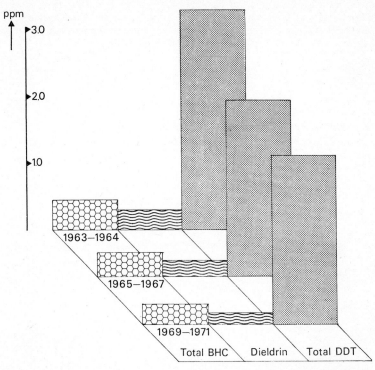

Fig. 2.5 Organo-chlorine pesticide residues in human fat, based on Table 12, p. 68, Report of the Government Chemist, 1971, Department of Trade and Industry, HMSO, London, 1972

REFERENCES

[1] The Haughley Experiment, a summary of the work carried out on the Haughley Research Farms, 1952–65, compiled by J. F. Ward, Haughley, England, 1966.

[2] (a) *Safe Use of Pesticides*, 20th Report of the WHO Expert Committee on Insecticides, Technical Report Series No. 513, WHO, Geneva, 1973.
(b) *Pesticides residues*, J. Thomson and D. C. Abbott, Royal Institute of Chemistry Lecture Series no. 3, 1966.

[3] *Pesticide Manual*, ed. H. Martin, British Crop Protection Council, Droitwich, Worcester, 1968.

[4] *Insecticides and Fungicides Handbook for Crop Protection*, ed. H. Martin for the British Crop Protection Council, 3rd edition, Blackwell Scientific Publications, Oxford and Edinburgh, 1969.

[5] *Further Review of Certain Persistent Organochlorine Pesticides used in Great Britain*, Report of the Advisory Committee on Pesticides and other Toxic Chemicals, HMSO, London, 1969.

[6] "Pesticide Residues in Foodstuffs in Great Britain: II Persistent organochlorine pesticide residues in selected foods", H. Egan *et al.*, *J. of the Science of Food and Agriculture*, 1966, **17**, 564.

[7] Pesticide Residues in the Total Diet in England and Wales, 1966–67:
(a) "Organisation of a Total Diet Study", J. M. Harries *et al.*, *J. of the Science of Food and Agriculture*, 1969, **20**, 242.
(b) "Organochlorine Pesticide Residues in the Total Diet", D. C. Abbott *et al.*, *J. of the Science of Food and Agriculture*, 1969, **20**, 245.

[8] *Persistent Pesticides in the Environment*, G. A. Edwards, C.R.C. Monoscience Series, Butterworths, London, 1970.

[9] Report of the Secretary's Commission on Pesticides and their Relationship to Environmental Health, US Department of Health, Education and Welfare, US Government Printing Office, Washington, DC, 1969 (referred to in the text as the Mrak Commission).

[10] "The Polychlorbiphenyls, their Occurrence and Significance: A Review", R. Edwards, *Chemistry and Industry*, 1971, 1340.

[11] (a) "Organophosphorus Pesticide Residues in the Total Diet", D. C. Abbott *et al.*, *Pestic. Sci.*, 1970, **1**, 10.
(b) "Pesticide Residues in Foodstuffs in Great Britain, V Malathion in Imported Cereals", E. G. Hill and R. H. Thompson, *J. of the Science of Food and Agriculture*, 1968, **19**, 119.

[12] "Toxicity and hazards to Man, Domestic Animals, and Wild Life from some commonly used Auxin Herbicides", J. M. Way, *Residue Reviews*, ed. F. A. Gunther, 1969, **26**, 37.

[13] *Pesticide Residues in Food*, Report of the 1971 Joint FAO/WHO Meeting, WHO Technical Report Series No. 502, WHO, Geneva, 1972.

Suggestions for further reading
Joint Survey of Pesticide Residues in Foodstuffs sold in England and Wales, 1 August 1967 to 31 July 1968 (second year), published by the Association of Public Analysts, London, 1971.
Pesticides: Benefits and Dangers, Symposium organised by Sir Wm. Slater, *Proceedings of the Royal Society*, Series B, 1967, **167**, 1007.
Silent Spring, Rachel Carson, Hamish Hamilton, London, 1962.

3

Agricultural pollution of food
Part II Miscellaneous problems

INTRODUCTION

THE PREVIOUS chapter has considered problems of food pollution
following upon the application of pesticide chemicals, either to protect
a crop growing in the field or after its harvest, in store, or in transit.
The applications in the field become increasingly necessary to cope
with the trend towards monoculture. They are also generally accepted
on the grounds that crops must be secured against the various pests, so
that food supplies can keep pace with the expanding population, and at
the same time improve the standards of nutrition in the developing
countries. Increasing productivity of the land is also required to com-
pensate for the diminishing areas of land available for cultivation,
especially in the western nations. Although more land could be made
available in some parts of the world by improved irrigation and drainage,
this is not likely to provide a short-term solution to the urgent need to
increase productivity. There is an unquestionable pressure on the
available land to increase its productivity; in addition to protecting the
crops by chemicals, yields can also be improved by the controlled
applications of chemical fertilisers. The applications of fertilisers without adequate controls, however, generate a number of environmental
problems and may also have an adverse effect on the quality of the soil.
This chapter (next section), however, is only concerned with the further
problems involving one of the mineral fertilisers, nitrate, which may be
carried over into the food consumed.

The trend towards intensive monocultivation systems of cropping is
also now being matched by the rapid development of intensive animal
rearing units. These involve the application of further chemicals, such
as antibiotics, to protect the animals in close confinement and also to
promote more rapid growth, and hormones also acting as growth promoters. The problems of residual contamination of animal produce by
antibiotics or hormones are considered in the third and fourth sections
of this chapter. A further problem derives from the practice of intensive
stock rearing and concerns the disposal of the animal sewage, which is

ed to a growing threat of pollution from heavy metals and patho-
strains of bacteria. The latter is dealt with in the fifth section.
le chapter is concluded by a final brief section on the problems of
cyclic hydrocarbons emitted from the exhausts of diesel-powered
tractors. The latter have also played a prominent part in the agricultural
revolution by increasing the efficiency of the industry, but may be
creating some minor problems of food contamination.

NITRATES

IT CANNOT be doubted that artificial fertilisers have been a major
factor in increasing crop yields approximately tenfold since the first
tentative applications began in Europe around 1750. The application
of artificial fertilisers is now carried out in most countries on a large
scale as illustrated by the world capacity for producing the three major
essential fertilisers, NPK:

1. 30 Mtons of nitrogen (N)
2. 18 Mtons of phosphorus (P), calculated as its pentoxide
3. 15 Mtons of potassium (K), calculated as its oxide.

The application of artificial fertilisers containing nitrogen is extremely
important and it has been estimated that the 30 Mtons applied to the
land represents an additional crop production equivalent to about 1000
Cals/day for each member of the earth's present population.[1] An appreciable fraction of nitrogen used by crops however still comes from more
traditional nitrogen sources.

Both phosphorus and potassium are available in natural mineral
deposits and supplies are judged to be perfectly adequate for the foreseeable future. Supplies of the mineral sources of nitrogen on the other
hand, such as Chilean saltpetre are extremely limited. There is an
unlimited reservoir of the element in the atmosphere but only a small
number of plants such as members of the pulse family, e.g. peas, beans,
with the co-operation of soil micro-organisms, are able to fix and to
utilise this atmospheric nitrogen. The great majority of plants require
their nitrogen in the form of a soluble compound such as nitrate. The
various forms of nitrogen in organic matter which are returned to the
soil from plant and animal wastes are partially converted into soluble
nitrates by certain other soil micro-organisms. A small quantity of
nitrate is also constantly reaching the soil as a result of lightning discharges in the atmosphere. The successful development of nitrogen
fertilisers has depended on chemical methods of fixing the nitrogen, the
most important process being the reaction with hydrogen to form
ammonia. The principal artificial nitrogen fertilisers are therefore

ammonium nitrate, ammonium sulphate and urea. The ammonium salts are also slowly converted into nitrates by micro-organisms in the soil and they can, therefore, make nitrogen available over a longer period of time than would be the case for the direct application of nitrates. The nitrogen cycle in nature is vital for the support of life and is illustrated in Fig. 3.1.

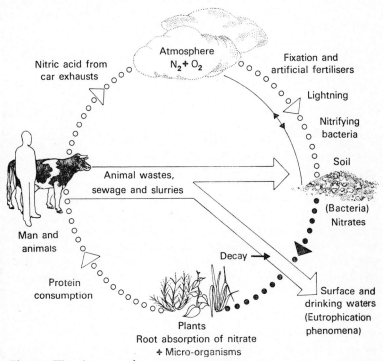

Fig. 3.1 The nitrogen cycle

Although it is beyond doubt that artificial fertilisers have increased crop yields, over-reliance on them at the expense of organic matter in the soil can create serious problems of soil texture. Further problems of a very different character arise because the amounts of artificial fertilisers applied to the soil are invariably greater than the immediate needs of the growing crops. Excess amounts may then be removed from the soils by rainwater and carried into streams and surface waters.

A similar problem, but extending over a rather longer period of time, may also arise when animal wastes from the intensive stock rearing units are spread on to the soil or on to pasture land. The three major plant nutrients then become enriched in the waters of freshwater lakes and land-locked seas where they have a similar effect on aquatic plants as on land-based plants stimulating, in some cases, explosive growths of algae and plankton. The average normal concentrations of the two more important elements in freshwaters are 0·06 ppm nitrogen and 0·005 ppm phosphorus, and it has been demonstrated that algal blooms are liable to occur if these two concentrations increase to 0·3 ppm and 0·01 ppm respectively. This phenomenon known as eutrophication leads to excessive oxygen demands in the water leading to anaerobic conditions and serious adverse effects on fish and animal life dependent on the aquatic environment. This may then create a serious ecological disturbance having an adverse effect on food supplies. The explosive growths of dinoflagellates which can occur occasionally in the early summer months, with one recent occurrence off the Northumberland coast, may also be due to excess nitrates in the sea-water. These organisms produce a highly toxic substance which may be absorbed by mussels and create a human health problem. All three essential nutrients are involved in eutrophication phenomena, but additional problems may arise from excess nitrate in surface waters extracted for drinking waters, or in the produce of the soils.

— The higher concentrations of nitrate in soils and surface waters due to agricultural applications are further increased by the emission of air pollutants from the internal combustion engine and from some industrial operations. The modern high compression internal combustion engine emits a number of gaseous products including some hydrocarbons (about 400 lb/1000 gallons fuel), carbon monoxide (about 3000 lb/1000 gallons fuel) and oxides of nitrogen (about 75 lb/1000 gallons fuel). In the United States alone, it has been estimated that about 20M tons of nitrogen oxides were discharged from vehicles and other stationary sources in 1968, so that the total global production of these oxides would have been in excess of the production capacity of industry for synthetic nitrogen fertilisers. The oxides of nitrogen after discharge are converted fairly rapidly into nitric acid by the combined effects of sunlight, oxygen and moisture in the atmosphere and, as a consequence, are doubling approximately the amount of nitrogen which is applied to the soil in the form of artificial fertilisers.

Plants absorb most of their nitrogen requirement from the soil in the form of soluble nitrate which is needed for the synthesis of a variety of proteins and other nitrogen compounds, and the absorption from a well-balanced soil is normally regulated to satisy the growth demands of the plant. Whenever there is an excess of available nitrate in the soil, the plant may accumulate more nitrate than it normally requires and some of this nitrate may remain in the plant tissues in the inorganic nitrate form. The problem has been reviewed by Walters[2] who reports that the normal nitrate content of a variety of plants ranges from 50 to

500 ppm nitrogen as nitrate, whereas plants grown on soils after intensive fertilisation by inorganic compounds of nitrogen may have concentrations more typically in the range of 2000 to 5000 ppm. The total nitrogen content of plants is of course considerably higher at an average level of about 30,000 ppm, but the major part of this nitrogen is in the form of protein and a number of other essential organic compounds. There is some evidence that these higher nitrate concentrations in herbage and animal feedingstuffs affect the health of livestock and there are certainly known to be human risks if the levels of nitrate in the diet or drinking water increase excessively. Any assessment of the nitrate problem at the present moment is made difficult by the lack of quantitative data relating the concentrations in man to the concentrations in the diet. A few instances of acute poisoning in young children have, however, been attributed to high levels of nitrate in occasional infant convenience foods such as samples of canned spinach.

Inorganic nitrate as such is not especially toxic to man. Dosages in the range of 0·3 to 1·0 g of nitrate have been used therapeutically to increase the acidity and volume of urine in adults without producing any noticeable ill-effects. The chemical transformation of nitrates, however, which may take place within the plant, in the food sources, or in man result in metabolites which are potentially a great deal more toxic to man. The first stage of the chemical changes normally leads to the formation of nitrite as an immediate reduction product:

$$NO_3^- \longrightarrow NO_2$$

$$\text{(nitrate)} \qquad\qquad \text{(nitrite)}$$

The reduction process is specially important when it occurs in the upper part of the alimentary tract and this happens only to a small extent in the case of the adult with a normal and fully developed digestive system; in the case of young children however the normal processes of digestion are limited by a lower stomach acidity; the micro-organisms which are responsible for the reduction are then able to establish themselves rather higher up the g.i. tract and absorption is therefore encouraged.

Inorganic nitrites, as such, are far from being a new factor in food and man, having found favour for the curing and preservation of certain meat products such as ham, bacon and luncheon meat, and also fish. Nitrites applied to fresh meat serve the twofold purpose of preserving the food from attack by micro-organisms and also of combining with the heme proteins to enhance and maintain the red colour. Organic nitrites have also been used medically in the past for the relief of high blood pressure.

The problems of excessive nitrate or nitrite levels are especially serious for young children as any nitrite after absorption is able to react with the haemoglobin of the red blood cells, with the formation of methaemoglobin. This substance is an oxidation product of the normal haemoglobin which no longer possesses the ability to fix oxygen and hence transport it to the tissues of the body. This results in a serious form of anaemia known as methaemoglobinaemia. There is also some further evidence from animal experiments that nitrite is also destructive of vitamin A and its precursor carotene, and also of vitamin E. In addition to acute forms of anaemia in young children, a further long term chronic mutagenic effect has also been confirmed which may take place at any time during the reproductive life-time of the individual. It has been shown that nitrite entering into the germ cells may interact with two of the four bases, adenine and cytosine, which go to make up the composition of DNA. A process of oxidative deamination takes place resulting in the substitution of an amino-group in the base by an oxygen atom. Adenine appears to be especially sensitive to this transformation.

$$\text{Adenine (A)} \xrightarrow{\text{NO}^-} \text{Hypoxanthine (H)}$$
$$\text{(6-amino-purine)} \qquad \text{(6-oxy-purine)}$$

This may then affect the normal coupling between the complementary pair of bases in the DNA chain resulting in a point mutation.

$$\begin{array}{c} A \\ :: \\ T \end{array} \longrightarrow \begin{array}{c} H \\ :: \\ T \end{array} \longrightarrow \begin{array}{c} H \\ :: \\ C \end{array} \longrightarrow \begin{array}{c} G \\ :: \\ C \end{array}$$

Although the mechanism of the mutagenic change is fairly well understood in this case, there are no data which are comparable with the data from radiation studies to relate the frequency of mutagenic changes by nitrite to the concentrations of either nitrate or nitrite in the human diet and no assessment of the magnitude of this effect is therefore possible.

The toxicity of nitrite may be even further enhanced by its known ability to react with certain amines, which are organic derivatives of ammonia, and may be present in the food or as metabolites of food in the body. The interaction in this case results in the formation of highly carcinogenic nitrosamines. An example of the nitrosation of an amine is the reaction with dimethylamine.

$$\begin{array}{c} CH_3 \\ \diagdown \\ NH + HNO_2 \longrightarrow \\ \diagup \\ CH_3 \end{array} \qquad \begin{array}{c} CH_3 \\ \diagdown \\ N-N=O + H_2O \\ \diagup \\ CH_3 \end{array}$$

This particular nitrosamine is formed on the scales of fish treated with nitrite and it has also been reported in samples of cured tobacco. The process of nitrosation has also been shown to take place in human gastric juices in *in vitro* experiments, and there is therefore a high probability that these compounds will be formed in the digestion of food. Compounds which are formed in this way are absorbed from the g.i. tract and subsequently transported in the bloodstream to give a fairly uniform distribution in the body tissues. They are known to be especially damaging to liver cells, producing a severe necrosis, and despite the fact that there is no evidence of their preferential concentration in the liver. Their carcinogenic nature is well established as they have been responsible for inducing malignant tumours in as many as eleven species of experimental animal, including the monkey. It has also been shown that the human liver can metabolise these compounds *in vitro* at rates comparable with those in experimental rats, although there is as yet no positive evidence which could identify liver tumours in man with the presence of nitrosamines. The experimental evidence, however, does suggest that man may be about as sensitive to nitrosamines as the experimental animals. The mechanism of this action almost certainly involves the methylating action of the compounds and of dimethyl-nitrosamine in particular, and their ability to transform another DNA base, guanine into methyl guanine. This change occurring in the somatic cells of the body is probably a factor in cancer production and the possibility of mutagenesis cannot also be ruled out. The biotransformations of nitrate in the foods consumed and in man and leading to products with increasing chronic toxicity requires some control in the agricultural applications of nitrate fertilisers in crop production and a constant surveillance of the dietary concentration levels. The World Health Organisation have recommended for adults an acceptable daily intake of 5 to 10 mg nitrate/kg of body weight, which is equivalent to about 420 ppm nitrate in the total diet of an adult person. The value recommended for young children is however about one-tenth of the adult value.[3] Although it is not possible to assess the current average daily nitrate intake from food sources some measure of control may be achieved by setting maximum concentrations of about 10 ppm in drinking water and of 500 ppm of sodium or potassium nitrate or 200 ppm of sodium or potassium nitrite in cured meats.

ANTIBIOTICS

A GREAT many chemo-therapeutic agents have been developed over the last thirty to forty years for the treatment of systemic infections due to pathological bacteria and other forms of micro-organisms. These agents include the synthetic sulphonamide group of drugs and the large group of antibiotics. A chemo-therapeutic agent is one which inhibits the multiplication of harmful micro-organisms and acts at concentrations which do negligible harm to the host. The antibiotics are a special group of these agents and have been defined as chemical substances produced wholly or partially by a special type of micro-organism, usually a fungus or bacterium, and which have the capability in dilute solution of inhibiting the growth of other micro-organisms.[4] They are produced, therefore, by the culturing of a range of specific micro-organisms, many of which occur naturally in soils. Special strains of bacteria which are frequently used for this purpose include the order of actinomycetes and the family streptomyces. Their presence in soils leads to the formation of some antibiotics at very low concentrations and the possibility of their natural presence in many of the plant foods and other produce consumed by man. A good example of a naturally occurring antibiotic is nisin which is produced by the bacterium streptococcus lactis and is found quite naturally in milk and in certain cheeses. Nisin is an example of a polypeptide type of compound with a similar composition to proteins and is consequently digested in a similar manner to other proteins without any harmful human effects at normal concentrations. Provided its concentration in the food consumed is not excessive, nisin is permitted also as a preservative for certain foods, but it has no medical applications. Nisin and other antibiotics may occur at very low concentrations in many natural products but the microbiologist has developed special processes for their culturing and isolation in much larger amounts.

One of the best known of the antibiotics is penicillin which is produced by a fungal mould and was discovered in 1929 by Fleming. Processes for its production and isolation on a large scale were developed by Florey and Chain at the start of the Second World War under the pressure of an anticipated increase in the demand for more chemotherapeutic agents. The outstanding properties of penicillin were quickly recognised from the time of its first clinical applications in 1941. Streptomycin, which is a metabolite of an actinomycete, was discovered in 1944 and subsequently more than one thousand antibiotics have been prepared and their properties studied. About one hundred of these have been well-characterised but only about thirty have been established successfully as therapeutic agents with a wide usage in modern medical practice. It has been estimated that the medical usage of the principal antibiotics in the UK in 1967 amounted to 185,000 kg (185 tonnes).[4]

The success of the antibiotics in human medicine led to a natural extension of their use to veterinary practice for the control of bacterial infections in animals. It became apparent quite rapidly that selected antibiotics were also effective as animal growth promoting agents. With the present trend towards intensive stock rearing involving the herding of large numbers of animals into confined spaces, two benefits were expected therefore to result from the addition of small amounts of an antibiotic to the animal feedstuffs, namely growth promotion and protection against infection. The antibiotics also provide some relief from the stresses imposed on animals in their unnatural surroundings. The total use of antibiotics in British agriculture including veterinary practice was estimated at approximately 110,000 kg (110 tonnes) in 1967.[4] About 50% of this amount is actually incorporated as a feed-additive. It is also known that a high proportion of all American livestock are fed regularly on medicated feeds.

Growth promotion is the stimulation of an animal's growth during the early years of its life resulting from the addition of small quantities of selected antibiotics to the feedstuffs. The precise mode of action of the growth promoters is not completely understood and certain other substances can also be used. These may include salts of copper and compounds of arsenic (see Chapter 5), synthetic hormones which are discussed in the next section, and a few other synthetic chemotherapeutic agents such as the nitrofurans. The practice of adding antibiotics to animal feeds varies widely from country to country. In the UK, mainly penicillin and the tetracyclines have been used with the occasional addition of others, such as streptomycin. These medicated feeds are especially advantageous in the rearing of young poultry and pigs, but in other countries the practice is also extended to calves, beef cattle and lambs. In British practice the concentration of the antibiotic in the feedstuff ranges from 20 to 100 ppm but the levels generally have tended to increase to concentrations from 100 to 200 ppm in order to achieve better control of stress and infections. A concentration of 20 ppm is equivalent to about 36 mg/day in the total feed consumed by a pig. The quantity administered in veterinary practice is, of course, much greater and may amount to an oral dose of about 1 g per treatment or an injection dose of 200 mg per treatment.

In addition to the general practice of dosing animal feeds, a number of antibiotics are also used for more specific purposes. The treatment of mastitis in the milking cow is one such example and may involve dosages of about 150 to 300 mg of penicillin into the udders of the animal. Several treatments at intervals of one to two days may be necessary, and the milk may be contaminated with penicillin residues for periods of up to 96 hours after the last treatment. A few other antibiotics, mainly tetracyclines, nisin and nystatin are used specifically for preserving and extending the shelf-life of fish and poultry and nisin may also be added to some processed cheeses and some canned foods. The practice of applying nystatin to control the growth of moulds on banana skins is also practised in some countries.

The growing practice of using antibiotics in agriculture and veterinary practice creates a twofold problem for man as illustrated in Fig. 3.2. The more serious of the two problems unquestionably concerns the

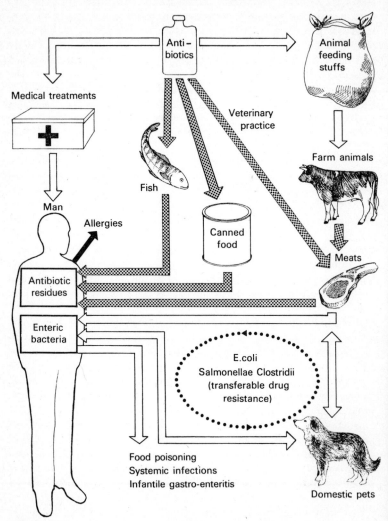

Fig. 3.2 Antibiotic residues and pathogenic micro-organisms in man

emergence of resistance to antibiotics among those enteric bacteria which are shared by man and his domestic animals and also the ability of certain strains to transfer their acquired resistance to others. This problem is reviewed in greater detail in the section on bacterial contamination later in this chapter. The less serious problem concerns the presence of antibiotic residues in foods and especially in animal produce. Very little is known about the concentrations in the foods consumed by man of either the natural antibiotics or those which are used in agricultural practice. There do not appear to have been any systematic surveys of antibiotic residues in foodstuffs, but it has generally been recognised that their applications in agriculture in accordance with approved procedures should not give rise to detectable residues in any of the animal foods consumed by man. It is not however possible to confirm their absence in foods owing to limitations of sensitivity in the best analytical procedures. These limits of detection for a number of the antibiotics in general use are as follows[5]:

Streptomycin	1·0 ppm in meat	0·2	ppm in milk
Penicillins	0·06 ppm	,,	0·006 ppm	,,
Tetracyclines	0·5 ppm	,,	0·1	ppm	,,

Where antibiotics are occasionally used for food preservation higher concentrations than these values may be encountered in the foods consumed, as for instance in the case of tetracyclines used in ice baths for the immersion of fish and poultry, when residual concentrations approaching 7 ppm may be reached. The tetracyclines are thermally unstable and concentrations are almost certainly reduced below 1 ppm after cooking. The maximum normal dietary concentration of all antibiotics is, therefore, likely to be well below 1 ppm, with a daily intake of much less than 1·5 mg which may be compared with a therapeutic dose amounting to a few hundred milligrams.

The fate of antibiotic residues entering the body from food is also largely unknown. The polypeptide antibiotics, nisin and bacitracin are almost certain to follow a similar ingestion pattern to proteins and to be broken down into their simpler component amino-acids prior to absorption. They should, therefore, have a low toxicity for man. With the possible exception of the streptomycin group, which include simple sugars in their composition, the other residues of antibiotics may well be absorbed through the alimentary tract relatively unchanged. Information about the subsequent biotransformations occurring in the liver or elsewhere in the body is again very meagre and the products of the biotransformations, whether active or inactive, are largely unknown.

The principal adverse human effects of antibiotics which have been identified are (a) allergic reactions, i.e. exceptional sensitivity of a few individuals to the exposure, (b) changes in the normal population of the micro-flora in the large intestine and (c) adverse changes in the stem cells of the bone marrow and subsequent haematological changes. A few individuals are exceptionally sensitive to antibiotic exposures and it is possible that this may be a consequence of a history of previous medical exposure. The allergy may vary from a comparatively mild reaction such as a skin rash to much more severe, and occasionally fatal, effects. The possibility must not be overlooked that exposure to antibiotics in food may also sensitise a person leading to complications in medical treatment involving penicillin and streptomycin if required subsequently. It hardly seems possible that antibiotic exposures from residues in food can produce other than minor changes in the micro-floral population of the large intestine. Such changes, if they were to occur on a larger scale, might lead to the invasion and establishment of alien strains of bacteria and possibly yeasts and fungi. There is also the possibility that bacteria establishing themselves in this way may already have acquired a resistance to the antibiotic and also an ability to transfer their resistance to other bacterial strains. This remains potentially one of the more serious consequences of antibiotic usage in farming and is discussed at greater length later in this chapter (pp. 96–8).

The majority of the antibiotics at the present concentrations in the food consumed can be expected to produce only mild effects in man. A few of the antibiotic residues in food may however give rise to more specific and well-defined effects. The tetracyclines for example are absorbed quite readily through the g.i. tract to be distributed throughout the body and deposited in the mineral parts of the bone and teeth. They appear to be strongly bound to calcium and may therefore inhibit the growth of fresh bone and teeth, with also some discoloration of the teeth. The effect may be more serious for growing children, and a temporary ADI of 0·15 mg/kg body weight has been recommended for them.

Chloramphenicol is another important antibiotic with more specific effects on the human organism. It is invaluable medically for controlling typhoid and paratyphoid in the human population, and although it is not used as a food additive, it may occasionally be used in veterinary practice for treating mastitis and other animal infections. Any residues which may find their way into the food consumed by man, are readily absorbed in the g.i. tract to become widely distributed in the body. They are especially damaging to the stem cells of the bone marrow and may cause a number of serious blood diseases. They may also be responsible for damage to the liver and to the eyes. It has therefore been strongly recommended that they should not be used for any purpose that might give rise to food residues.[5]

The British Joint Committee on the Use of Antibiotics in Animal Husbandry and Veterinary Medicine has referred briefly to the problem of residues in the food consumed, and considered that residues are likely to be present in some of these foods. These amounts are expected to be low and perhaps below the limits of detection by the analytical methods available, and the Committee considers that there is no evidence that they have caused any harm to man with the possible exception of occasional residues in milk. The Committee has however made a firm recommendation for a survey to be carried out to determine the presence or the absence of antibiotic residues including their degradation products in animal produce consumed in the UK.[4]

HORMONE RESIDUES

THE TERM "hormone" is derived from a Greek word meaning to stir up or to excite. Hormones are, therefore, stimulating agents and are organic chemical substances which are either polypeptides, such as the well-known insulin, or steroids, such as the various sex hormones. They are secreted by special tissues in the body, normally the endocrine of ductless glands. Insulin is produced, for instance, in the islet cells of the pancreas and the sex hormones in the gonads. The hormones secreted by the various glands are transported in the bloodstream and have the ability to alter the rates at which reactions mediated by the enzymes are taking place in the specific cells and tissues of the body. The effect can be twofold in that the normal reaction can either be speeded up appreciably or almost arrested. Hormones are, therefore, present quite naturally in the animal body and are also quite widely distributed in the plant kingdom. An example of a naturally occurring hormone having similar properties to the female sex hormone is produced as a metabolite by a fungus, Fusarium graminearium, which is liable to grow on cereal grains stored under damp conditions. In the case of animals, their well-being depends to a very large degree on a delicate balance of controls exercised by hormonal activity. The gradual isolation, identification and characterisation of many hormones since the original discoveries at the start of the present century has led, in turn, to the production of a number of synthetic sex hormones, notably diethyl stilboestrol (DES) and hexoestrol, which are possible additional hormone contaminants in the human diet. These artificial sex hormones are of some concern when problems of human food contamination are involved.

The natural sex hormones are of two types, the male sex hormones or androgens, and the female sex hormones, or oestrogens. Both types of hormone affect the development of secondary sex characteristics and the latter, in addition, regulate the female menstrual and reproductive cycles. At the appropriate stage of development, the gonads are stimulated to produce the sex hormones by other hormones, the gonadotrophins, which are secreted by the hypothysis or pituitary gland at the base of the skull and are circulated in the blood. The female sex hormones are also known to have a general effect on female metabolism and to be effective in promoting the growth of young animals. They are found to affect not only the efficiency and the rate at which the food proteins are converted into animal protein, but also to promote a better and more acceptable distribution of fat in poultry birds.

The synthetic hormone DES has found special favour as a growth promoter in animal feeds especially in the USA where it has been estimated that the compound is mixed with the feeds of nearly three-quarters of the cattle slaughtered annually in that country. In the UK its use appears to have been limited to veal and poultry production; in the latter case, the compound is in pellet form and is implanted in the neck of the bird when it acts as a chemical caponising agent. It is not used, however, in broiler production, which accounts for about 85% of the poultry meat produced in the UK. DES has all the properties of a natural oestrogen but with about three to five times the activity of a natural hormone such as oestrone. It is a fairly simple type of chemical compound having the following formula, and some similarity with the formulae for the DDT compounds and the synthetic plant hormones, 2, 4-D and 2, 4, 5, -T.

The natural oestrogens are steroids having a far more complicated ring structure, the compound oestrone, for instance, having the same basic structure (left) as the well-known drug cortisone (right).

The concentrations of any natural hormones in food are extremely small and below the limits of detection of the most sensitive analytical procedures available. There are also no systematic data available to enable any estimate to be made of the average concentrations of synthetic hormones in animal produce, although these again should be extremely small if the proper recommended procedures and routines are followed in the feeding of the animals. It is stipulated that hormone supplemented feeds should be discontinued at least seven days prior to slaughter and the heads of birds treated with the pellets should be discarded. The hormone is normally quite rapidly assimilated and eliminated by the animal, most of it over a period of twelve hours with only small residues at this stage remaining in the liver and possibly the kidneys. Recent data from the United States for June 1972, however, show that the compound DES was detected in about 5 % of the animal livers tested. Some of these residual levels may have been due to failure on the part of the stock rearer to observe the requisite period of hormone-free feeds. In these circumstances the assessment of any average concentrations in meat products will not be possible. Systematic data for foods and residual levels in man are, therefore, not available but it is practically certain that any concentrations will be extremely small and below the limits of detection of existing analytical procedures.

The effects of DES on experimental animals have been investigated and in the case of rats it has been shown that the ingested compound is rapidly eliminated, partly in the urine and partly in the faeces. It appears to be fairly rapidly absorbed through the g.i. tract, combining in the liver with glucuronic acid to form a glucuronide which is then partly eliminated by the kidneys. The remainder re-enters the g.i. tract in the bile secretion from the liver and is partly eliminated in the faeces and partly recirculated. Higher concentrations of DES and longer periods of retention may, therefore, be expected in the liver and kidney tissues.

Human concentrations of DES or any other hormone from food are most unlikely ever to reach the values where the metabolic changes associated with the oestrogenic activity of hormones would become apparent. With DES, however, there is serious concern about its carcinogenic properties, which have been confirmed in experimental mice at concentrations in the diet of a few ppb, below the limits of analytical measurement. The same hormone has also been implicated in the USA in a number of uterine cancers which have developed in women between the ages of 15 and 20 years and whose mothers had been treated with DES tablets to prevent a threatened abortion. A Chicago gynaecologist is reported to be currently trying to track down about 1600 women who had participated in these treatments about ten years ago. There seems to be little doubt, therefore, that DES is a highly carcinogenic substance for animals and that it has also been the cause of some human cancers.

For this reason, its use has been totally banned in Sweden and West Germany and is under close scrutiny in the United States.

The problems of hormone residues in food and especially those of DES have been closely examined in the USA and were the subject of a recent report published by a task force of the US Centre for the Study of Responsive Law[6] from which some of the information summarised in this section has been obtained.

BACTERIAL CONTAMINATION

THE CHANGES in agricultural practice in Britain and elsewhere with increasing emphasis on intensive cultivation of the soil and intensive rearing of animals in confinement have already been referred to. These changes are already responsible for a variety of fresh polluting agents in food, but perhaps no problem is potentially more serious than the increasing danger of food poisoning due to pathogenic micro-organisms which are mainly found in animal produce. Animal foods have always been a major source of food poisoning incidents, being responsible for about 80% of outbreaks in recent years[7] and there are fears that the total number of such incidents especially those due to salmonellae micro-organisms, may be increasing.

The problem is illustrated by two outbreaks of food poisoning caused by the bacterium Salmonella virchow, both of which occurred in 1968.[8] The earlier of these was centred around a suburban lawn tennis club in Liverpool at the end of June and involved about 120 people who had all eaten spit roasted chicken. Altogether some 160 infections in the Liverpool area from mid-June to the end of August 1968 were subsequently attributed to a similar source of food. The outbreak at the tennis club was fairly severe, requiring hospital treatment for eleven members of the club. The second and later outbreak occurred in October in a new special-care baby unit in a West Midlands hospital. This unit normally accommodated between fourteen and twenty-one babies and ten cases of severe diarrhoea attributed to poisoning by the same bacterium occurred in a period of about a fortnight. The source of the epidemic was traced to one of the mothers who had earlier consumed roast spit chicken which had originated from the same supplier of chickens involved in the Merseyside outbreak.

Although both these outbreaks had a common origin, the responsible agent, the Salmonella virchow was transmitted in two very different ways. In the Merseyside outbreak, the food consumed by the members of the club was the source of infection in every case, whereas at the hospital cross-infection was the factor in the spread of the poison. In the former case, although the primary infection occurred elsewhere, the spread of the contamination via the food was due to the unhygienic conditions prevailing in the kitchen of a food establishment, and was also partly due to the failure to follow the proper procedure in the preparation and cooking of the chickens. The chickens had been kept in deep freeze overnight but only one to two hours were allowed for thawing out before roasting. This was not an adequate time to allow for the complete thawing out of the interior of the carcases so that during cooking the temperature rise in the interior was insufficient to destroy the bacteria which could have been completely killed off at a high enough temperature. As a much longer thawing out time is necessary to ensure satisfactory cooking of the inside of the carcases, the spit itself became contaminated as were the utensils and instruments used in the quartering of the cooked chickens prior to packing.

The common factor in both these outbreaks was frozen chickens from a Cheshire packing station. This station was the centre of a farming co-operative with sixteen rearing farms receiving day-old chicks from the station, the chicks being kept on deep litter for about nine weeks. Samples of dressed chickens at the packing station gave positive results for Salmonellae despite very satisfactory standards of hygiene. Eight out of twelve rearing farms which were also tested were found to have infected birds, but it was not possible to establish with any certainty the primary source of the infection, which may have been due to a human carrier or to the poultry protein feed supplements, consisting of meat, bone meal or poultry offal meal.

A number of disturbing features were revealed by the investigations into these two outbreaks.

1. The emergence of Salmonella virchow as a source of food poisoning, especially subsequent to 1967. In 1966, for instance, only two out of 2500 cases of Salmonella food poisoning were attributed to this particular strain.
2. The spread of infection to dairy cows grazing on land manured by litter from two of the rearing farms.
3. The tendency of medical treatment by means of antibiotics to prolong the effects of poisoning among some of the more seriously affected Merseyside patients in hospital.

The problems associated with 1 and 3 may well have a common link with the use of antibiotics in animal feeds, although the original papers do not indicate whether medicated feeds were in use at any of the sixteen rearing farms. These problems which may be associated with the emergence of bacterial resistance to antibiotics are discussed later in this section.

The disposal of infected litter by spreading it on to grazing land highlights a further problem involving the spread of infection to livestock in the fields and, thence, to man, either by contact with the animals or through the consumption of infected milk. This problem is becoming increasingly serious as the trend towards the intensive system of animal production continues.[9a] These intensive systems are assuming more and more, the characteristics of an industrial operation and have been described as "factory farming". The operation involves the "convenience feeding" of large numbers of animals at a high population density such as 50 to 150 sq. ft. per animal in the case of cattle. The operations involve the farm use of the medicated feedstuffs described earlier and the production locally of large volumes of animal wastes for disposal. Tables 3.1 and 3.2 provide a general guide to the average

Table 3.1

Animal waste production[9b]

Farm animal	Estimates of average waste production lb/day per animal		
	Wet	Dry solids	Nitrogen
Cattle and calves	60	9	0·2
Pigs	17	3	0·07
Poultry	0·4	0·1	0·0025
Sheep	7	2	–

Table 3.2

Total animal waste production per annum[9c]

Farm animal	Est. no. in 1968 (millions)	Total waste production Mtonnes/year	
		Dry solids	Nitrogen
1. Cattle and calves	12	18·0	0·4
2. Pigs	7	3·5	0·08
3. Poultry	127	2·2	0·06
Total 1+2+3	–	23·7	0·54
4. Sheep	28	9·3	0·26
Grand total 1+2+3+4	–	34	0·8
cf. Human wastes	–	5	0·2
Rainfall deposition	–	–	0·25

quantities of total waste produced. The estimate for total waste production expressed as dry solids, mainly organic matter, amounts to about 34 Mtonnes per annum. This is about seven times the amount of dry solids in the whole of the human sewage produced in the UK. The total quantity of nitrogen in the animal wastes is estimated to be about four times the quantity produced in the human wastes. Not all of the animal waste is produced at the moment in intensive units, but the disposal and pollution problems will grow increasingly as the change-over from conventional husbandry, involving small units, to the large intensive units gains momentum. The total number of farm animals is also increasing and the problem is further aggravated by the liquid form of the waste from the intensive units.

A number of problems arise from these changes. In the first instance, any organic matter which may find its way into surface waters is oxidised by micro-organisms, creating a demand for dissolved oxygen (expressed as the biological or biochemical oxygen demand, BOD). This leads to the depletion of the oxygen content of the water so that it is no longer capable of sustaining the normal level of fish and animal life. The wastes are also relatively rich in two essential nutrients, nitrogen and phosphorus, so that the productivity of the water for supporting plant life is increased and explosive growths of algae and plankton may occur. Their subsequent decay and oxidation also leads to further severe oxygen depletion in the water, with even greater destruction of the normal fish and animal life of the aquatic environment. The discharge of animal wastes reaching surface waters may, therefore, result in a depletion of some food supplies, whether in a sea water or a fresh water environment. The BOD from all the animal wastes has been estimated to be approximately five times the corresponding BOD from all human sewage produced in the UK if this waste should reach surface waters in an untreated form.

These problems are specially acute around the large intensive stock farms with a very large bulk of mainly liquid manure requiring disposal. They are aggravated by the difficulties in treating the wastes satisfactorily at sewage farms and their unsuitability for direct spreading on the land as was customary with the more solid forms of animal waste in the past. This is due to a large extent to the offensive and persistent odours of the waste. Ideally, animal wastes should be returned to the soil and procedures for doing this are under investigation at present, but a number of problems are likely to remain due to the composition of the wastes and the presence of a number of potential contaminants finding their way from the soil into plant produce. These problems, which are also under investigation, may be summarised as follows:

1. The tendency for the wastes to have relatively high concentrations of certain trace elements, more especially of copper and zinc, and possibly of nickel. The presence of copper is likely to cause special concern when the wastes from pig-rearing units are disposed of as copper salts are frequently incorporated in the pig feeds. The copper concentrations might be sufficiently high to prove harmful even to growing crops and also to cattle feeding on the treated herbage. The problems of trace mineral elements which arise even

more acutely from industrial operations are discussed further in Chapter 5.

2. The relatively high nitrogen and phosphorus concentrations in the wastes aggravating the problems discussed in the section on mineral fertilisers and above.

3. The dispersal of pathogenic enterobacteria in the environment as happened with the litter from the chicken rearing units described on page 93.

Large numbers of bacteria and other micro-organisms populate the alimentary tract of man and his domestic animals. The vast majority of these are coliform organisms with rather different strains of these organisms occurring in the bowels of animals from those in man. They are generally harmless to man, although more virulent strains of, for instance, Escherichia coli, sometimes produce toxic metabolites and, as such, they were implicated in the serious outbreaks of infantile gastro-enteritis which occurred on Teesside at the end of 1967 and in Manchester at the end of 1968. Another strain of the same organism may also be responsible for some infections of the urinary tract in man. The proliferation of coliform organisms in man and animals facilitates their detection, so that their concentrations in samples of drinking water provide early presumptive evidence for bacterial pollution from sewage. Some other strains of bacteria are also shared by man and his domestic animals, including a number of pathogenic strains of Salmonellae. Altogether about 1000 species or serotypes of these bacteria have been reported and these include Salmonella typhi, Salmonella typhimurium and Salmonella virchow. The first of these, which is the typhoid organism, although found in animals, is only infective for man, whereas the second, typhimurium, infects a variety of animal hosts being a common cause of food poisoning in man and sickness in animals. The emergence of Salmonella virchow as an infective food poisoning agent for man has already been referred to in this section. All these, and other organisms, such as those responsible for tuberculosis and brucellosis, are transmissible from animals to man, either indirectly through the contamination of animal produce consumed as food or, less frequently, by direct contact with the infected animal.

The risks of human infection by pathogenic organisms in food have been aggravated by two additional factors both resulting from the use of antibiotics in agriculture, either as feed additives or in veterinary practice. These two factors are (a) the emergence of mutant strains of micro-organism which possess resistance to a particular antibiotic or multiple resistance to several antibiotics, and (b) infectious, or transferable, drug resistance between different strains of bacteria. Both of these effects appear most readily among the enterobacteria and complicate the medical treatment of infections involving these micro-organisms; they may create special medical problems in the case of the antibiotic treatment of systemic fevers such as typhoid.

The earliest suspicions that the first of these factors might be operating in human bacterial infections was aroused soon after the introduction of antibiotics into animal feeds in the early 1950s. Special attention was drawn to these problems however in Japan in 1960, when the treatment of outbreaks of bacillary dysentery proved extremely difficult to control owing to the emergence of multiple drug resistance among the responsible shigella organisms. These bacteria were found to be resistant to streptomycin, chloramphenicol, tetracyclines and sulphonamides, and the resistance was also found to be transmissible to other infective bacteria. Evidence of resistance to antibiotics was also found in the UK in 1965 in the Salmonella typhimurium responsible for infective diarrhoea among calves and also responsible for a small number of fatalities in human cases of poisoning attributed to the same organism. The strain of Escherichia coli in the Teesside epidemic of infantile gastro-enteritis and responsible for ten deaths was also difficult to treat because of multiple drug resistance, although the original source of the infection could not be positively identified in this case. Growing concern with these and other outbreaks was responsible for the setting up of the Joint Committee on the Use of Antibiotics in Animal Husbandry and Veterinary Practice under the Chairmanship of Professor Swann in 1968.[4]

The emergence of resistant bacterial strains cannot have been entirely unforeseen as certain species are known to possess an innate resistance to certain antibiotics. Penicillin, for instance, is described as a narrow spectrum antibiotic, being particularly effective on Streptococci and Staphylococci, but having little effect on Salmonellae. The tetracyclines, by contrast, are broad spectrum antibiotics, being effective against a wide range of bacteria. Development of antibiotic resistance is a natural evolutionary process as it provides a selective advantage to any mutants which acquire the resistance factor. The precise mechanisms of acquiring this resistance may not always be clearly understood but, in some cases, they may involve changes in the genetic coding of the cell, enabling it to produce certain enzymes which can transform the antibiotic into harmless metabolites. This is the case for some Staphylococci which are able to produce an enzyme, penicillinase, which can transform and inactivate penicillin. Thus, bacteria which have acquired this favourable mutation can prosper in an otherwise hostile environment and by an extension of the same process may develop their resistance to other antibiotics with the acquisition of multiple resistance. The development of these resistant strains is especially favoured by continuing exposure to low concentrations of the antibiotics, such as those which are found in animal feeds. Therapeutic doses of antibiotics, by comparison, are more likely to lead to almost total destruction of the bacterial populations. The acquisition of drug resistance appears to have no effect on the virulence of the bacteria, and its ability to harm the host animal.

The emergence of the second factor, "transferable drug resistance", was far more unexpected. This factor involves the ability of a particular strain which has acquired the resistance factor to act as a donor and transfer its acquired resistance factor to other strains acting as recipients. The recipient strains may not have been exposed to any concentration of the antibiotic at any time. It also appears to have been established that a donor strain which has been exposed only to a single antibiotic may also have acquired multiple resistance with the ability to donate this multiple resistance to other strains. Most of the evidence for transferable drug resistance is based on *in vitro* experiments; and the occurrence of *in vivo* transfer is generally inferred from epidemiological investigations. The Swann Committee concluded that "there are strong indications in the enterobacteriaciae, at least, that the prevalence of multiple antibiotic resistant strains has increased considerably over the last decade or so, and that most of this resistance is transferable".[4]

The Committee is also satisfied that the administration of antibiotics to animals, either in feed lots or in veterinary treatment, has also caused some harm to human health as a result of the emergence of bacterial drug resistance. It is just possible that some of this resistance may have been acquired as a result of medical practice, as antibiotics have also been used in medical therapy on a large scale since the end of the Second World War. In practice, it would not be possible to distinguish resistance factors emerging from applications to agriculture or to medicine, but it is probable that a major contribution is due to agricultural practice owing to the continuous administration of the antibiotics at low concentrations in the feed lots. The bacterial contamination of animal produce by resistant strains will, therefore, be a major factor in human contamination. The contribution resulting from the disposal of animal wastes on to the soil or into surface waters will almost certainly be a much smaller one. The widespread dispersal of drug-resistant coliform organisms is illustrated by the data in Table 3.3 for samples of river water and one sample of sea-water having a relatively high concentration. It is notable that the heaviest concentrations were found in the river waters at, or after passing through, large urban areas and that the lowest concentrations were in the predominantly rural areas. This suggests that human sewage is still a major environmental source of the resistant coliform organisms, and that the pollution from agricultural wastes is at least under control at the present time. The coliform resistance to chloramphenicol in some of the samples examined was found to be transferable to Salmonellae typhi, and this is clearly a highly undesir-

Table 3.3

Some concentrations of antibiotic-resistant coliform organisms in specimens of river water[10]

Source	Number of viable coliform organisms per ml Numbers resistant to				
	Total	Chloramphenicol	Tetracyclines	Streptomycin	Ampicillin
Tame, after Birmingham	5000	30	200	80	200
Cole, after Birmingham	5000	30	100	80	400
Ribble, at Preston	2000	16	50	25	250
Avon, after Bristol	75,000	10	150	200	5000
Avon, before Bristol	150	0	0	4	40
Severn, before Stourport	50	0·15	2	8	30
Great Ouse, before Huntingdon	150	0	0	0	0
Jed, at Jedburgh	50	0	0	0	0
Cam, at Cambridge	50	0	0	0	20
Sea water from bathing beach	200	0·25	2·5	2·5	25

able situation as this antibiotic is the best drug available for the treatment of typhoid and paratyphoid.[10]

It has been found that the micro-organisms can persist for periods of up to twelve weeks in the animal waste slurries. There have only been a limited number of investigations into the fate of the micro-organisms once the slurries, either before or after treatment, have been applied to the soil or to pasture land. A strain of Salmonellae has been found to persist up to 12 weeks when the slurry was applied directly to the soil. The soil was found to be fairly effective in removing the bacteria from the liquid slurry, probably by absorption on to a clay mineral component of the soil. The micro-organisms are, therefore, fairly readily removed from any slurry which is percolating through the soil so that there should be very little entry into surface waters. The bacteria in the soil find themselves in competition with the normal soil bacterial populations and tend to die off as a result. Other environmental factors, such as the amount of sunlight, temperature, rainfall and the extent of grass cover, also influence the persistence of the organisms in the soil. The bacteria in the top layers of pasture for example have much shorter life-times of seven to eight days, but they may be a source of infection for grazing animals depending on their concentrations in the slurry; concentrations of about 10^6/ml of Salmonellae applied to pasture were found to infect calves in a few days but there was no infection when the concentrations were reduced one thousandfold.[9d] Although the possibility of food chains leading to the contamination of plant and animal produce, from the application of slurries to the land exists, there is no evidence that this is a principal source of human contamination by micro-organisms.

Contamination of human beings might be expected from the animal food products originating from the intensive stock rearing units. In 1967, in England and Wales there were 4256 incidents and 11,095 cases of food poisoning, of which 3259 incidents and 5527 cases were attributable to Salmonellae organisms.[7b] There could also have been a large number of unrecorded cases. Animal produce was found to be a major source of these infections and Salmonellae typhimurium were responsible for 50% of all the outbreaks where a cause could be definitely established. It has been reported that about 1·5% of all samples of butchers' meat are infected to some extent with Salmonellae, but most of these should be destroyed in cooking. It is not possible however at the moment to establish how much of this contamination by Salmonellae is attributable directly to the intensive stock rearing of the animals.

The presence of Salmonellae organisms in a fair number of samples of butchers' meat requires exceptional standards of hygiene when the meat is handled in the kitchen and in catering establishments. Although the organisms may be largely destroyed in the cooking process, any kitchen surfaces, utensils and instruments used in the preparation of the meats prior to cooki ng may harbour the contamination, which may then be passed on to oth er foods if these surfaces have not been properly cleaned after handling t he meat. The position has been well expressed in the report of the Swa nn Committee. "At one end of these chains of transmission is an infec ted animal; at the other end the infected food reaches man. Even if th e food itself is sufficiently cooked to kill Salmon-ellae organisms, there are many w ays in which infected food may contaminate hands, cutlery and kitchen sur faces and the infection may then reach the consumer through the cont amination of uncooked or previously cooked dishes."[4] The trend towards intensive stock rearing which may continue to gain momentum, coupled with the use of feeds containing antibiotics must, therefore, compel very much higher standards of hygiene in the food industries, catering establishments and in the home kitchen.

Micro-biological standards of food quality are difficult to formulate, mainly because of the practical problems of sampling bulk supplies of food. Freedom from pathogenic organisms is obviously a desirable criterion and attempts have been made to relate the presence of patho-gens to total viable bacterial counts and to coliform counts. A large retail organisation in the United Kingdom has laid down the standards for meat produce as shown in Table 3.4.

Table 3.4

Retail bacteriological standards (after 24 hours at 65–70^{0}F)[11]

Sample	Total count of micro-organisms, 48 hours at 30^{0}C	Coliform organisms	Salmonellae
Sausages	< 500,000/g	< 500/g	Absent in 25 g
Meat Pies	< 200/g	Absent in 0·1 g	Absent in 50 g

The same organisation recommends that coliform organisms should be absent in 1 ml of fresh cream, 0·1 ml of yoghurt and 0·1 g of soft cheese, examined either immediately or after overnight storage at 40^{0}F.

POLYCYCLIC HYDROCARBONS IN DIESEL ENGINE FUMES

THE EXHAUSTS from diesel engines can be hazardous, either when inhaled or when certain contaminants are transferred to food. The exhaust gases contain small amounts of the highly carcinogenic poly-cyclic hydrocarbons such as 3, 4-benzopyrene (*left*) and dibenzanthra-cene (*right*).

The hydrocarbons are usually emitted in the form of sub-microscopic particles about 1 μm in size. These extremely fine particles may remain suspended in the atmosphere for long periods of time or they may attach themselves to larger smoke particles, greater than 50 μm in size, and then fall fairly rapidly to the ground. An idling diesel engine may emit about 1·7 mg of polycyclic hydrocarbons per minute, which is about four times the quantity which would be inhaled by smoking 500 cigarettes. A petrol engine also emits these substances but normally in very much smaller amounts. These hydrocarbons emitted from the exhaust of the diesel engine will be considerably diluted in the atmosphere, although the hazard to the lung is almost certainly the more serious of the two problems produced by these compounds.

The contamination of food will result from the deposition of the particles eventually on to soil or on to crops growing on the soil or into surface waters. It is appropriate to consider the problem in the chapter on agricultural pollution, mainly because of the almost complete change-over to diesel powered farm tractors in the last 15 to 20 years. A great deal of farm working also involves idling or very low loading of the engine and these are the conditions which favour the maximum formation of the hydrocarbons. In the modern high compression diesel engine, a peak firing temperature of about 5000°F and peak pressures of about 1 ton/in² may be reached for about 10 msecs. The fuel is injected as a fine spray and the simpler hydrocarbons in the liquid particles which are not fully aerated at the time and hence fail to make adequate contact with oxygen at the firing stage are then exposed to conditions of temperature and pressure which favour their cyclisation. It has been estimated that more than double the amount of fuel is used to keep the engine running smoothly than is required in theory.[12] Estimates of the total amounts of polycyclic hydrocarbons produced by farm tractors and other forms of transport in rural areas are difficult to make and there do not appear to have been any systematic investigations of these compounds in the environment leading to the contamination of food or drinking water. It seems most unlikely that there will be any absorption of these compounds from the soil into plants, such as cereals, salads and orchard produce, and the contamination problem will concern only fresh produce exposed to the fallout of the particles.

The possibility of these compounds being present in agricultural produce has to be considered in relation to the certainty of small concentrations being present in a number of foods such as fish and meats which have been preserved by the traditional method of smoking. A number of these compounds are certainly formed in the combustion of the lignin compounds in wood whenever the temperature exceeds 350°C and as this is one of the oldest forms of food preservation with the additional attraction that it also enhances colour and flavour, human exposure to small amounts of these compounds has been happening for some time. The problem is special to those countries which rely heavily on smoke-preserved foods. Their presence in smoked foods from Iceland has been reported (Table 3.5).

Table 3.5

Concentrations of benzopyrene in smoked foods[13]

Food	Concentrations of benzopyrene ppm
Mutton	1·3
Cod	0·5
Trout	21·0

Benzopyrene has also been reported at concentrations of about 1 ppb in some samples of smoked haddock and salmon and at concentrations between 5 and 10 ppb in charcoal broiled meat.[13] The actual concentrations of these hydrocarbons in fresh agricultural produce are likely to be well below 1 ppb and it seems unlikely therefore that the maximum dietary intake could ever exceed 1 to 2 μg.

The risk of oil pollution of the seas, which is increasing almost daily, may be an additional source of the polycyclic hydrocarbons through their contamination of marine produce. The Government Chemist has recently extended the analyses of food to include some of these compounds in fish and shellfish from coastal waters. All the samples of fish which were tested were found to have concentrations below the limits of detection of the analytical method (less than 0·003 ppb), but shellfish were found to contain very low concentrations of benzopyrene ranging from 4 to 16 ppb with smaller concentrations of some other polycyclics.[14]

Any of these compounds which are ingested from food are unlikely to be absorbed at all readily through the alimentary tract and should tend to pass straight through the digestive system with elimination in the faeces. The gastro-intestinal tract may be expected, therefore, to be the critical organ of the body to exposure from these particular compounds. This particular risk is illustrated by the high incidence of gastric carcinomas in rural areas in Iceland, this form of cancer being responsible for about 35% of the deaths in some rural populations.[15]

At the normal concentrations which have been found in food samples however, it appears most unlikely that despite their highly carcinogenic character, there is at present any serious health risk to the general population from food sources of these compounds.

REFERENCES

[1] The Fertiliser Industry, The Castner Memorial Lecture, B. Timm, *Chemistry and Industry*, 1970, 1518.

[2] *Nitrate in Soils and Plants and Animals*, H. Walters, Blackfriars Press, Leicester, for the Soil Association, Haughley, Suffolk, England.

[3] *Evaluation of Food Additives*, WHO Technical Report Series No. 488, WHO Geneva, 1972.

[4] Report of the Joint Committee on the Use of Antibiotics in Animal Husbandry and Veterinary Medicine (referred to as the Swann Committee after its Chairman), Cmnd 4190, HMSO, London, 1969.

[5] *Specifications for the Identification and Purity of Food Additives and their Toxicological Evaluation : some Antibiotics*, 12th Report of the Joint FAO/WHO Expert Committee on Food Additives, WHO Technical Report Series No. 430, WHO Geneva, 1969.

[6] *Sowing the Wind*, Harrison Welford, US Centre for the Study of Responsive Law, discussed in *Nature*, 1971, **235**, 216 (see also *Nature*, 1972, **238**, 67).

[7] (a) "The Spread and Control of Water and Food borne Infection", R. Cruikshank, chap. 15 in *The Theory and Practice of Public Health*, ed. W. Hobson, Oxford University Press, London, 3rd edition, 1969.

(b) *Bacterial Food Poisoning*, introduced by Sir James Howie, Royal Society of Health, London, 1969.

[8] (a) "Food Poisoning due to Salmonella Virchow", Professor A. B. Semple *et al.*, *British Medical Journal*, 1968, **4**, 801.

(b) "Outbreak of Gastro-enteritis due to Salmonella virchow in a Maternity Hospital", B. Rowe et al., *British Medical Journal*, 1969, **3**, 561.

[9] (a) Farm waste disposal in the UK, Ministry of Agriculture, Fisheries and Food, Short-term Leaflet (67), London, 1969 (revised 1973).

(b) "Changing Practices in Agriculture and their Effects on the Environment", R. C. Loehr and S. A. Hart, *Critical Reviews in Environmental Control*, 1970, 69.

(c) "Effects of Water Pollution by Animal Effluent", M. Owens, *J. of the Science of Food and Agriculture*, 1972, **23**, 793.

(d) "Disease Hazards of Pathogenic Organisms in Farm Waste Material", D. A. Haig, *J. of the Science of Food and Agriculture*, 1972, **23** (6), 795.

[10] "Incidence in River Water of Escherichia Coli Containing R Factors", H. Williams Smith, *Nature*, 1970, **228**, 1286, also *Nature*, 1971, **234**, 155.

[11] "Bacteriological Standards for Foods: a retailer's point of view", R. L. Stephens, *Chemistry and Industry*, 1970, 220.

[12] *Diesel Fumes : an increasing hazard to health*, J. A. Butler, published privately at Sunnymead, Winchester Road, Waltham Chase, Southampton.

[13] "Polycyclic Hydrocarbons in Icelandic Smoked Food", E. J. Bailey and N. Dungal, *British J. of Cancer*, 1958, **12**, 341.

[14] Report of the Government Chemist, 1971, Department of Trade and Industry, HMSO, London, 1972, p. 38.

[15] "The Special Problem of Stomach Cancer in Iceland", N. Dungal, *J. American Medical Association*, 1961, **178**, 789.

4

Radioactive pollution of food

INTRODUCTION

RADIOACTIVE POLLUTION of the environment resulting from human activities is very nearly contemporaneous with the pollution from pesticide chemicals. Both of them became serious sources of environmental pollution after the Second World War, and as a result of discoveries before the start of the war and of large-scale developments during the war years. There is a further parallel between these two major classes of environmental pollutant, both of them occurring naturally in certain forms although awareness of their presence in nature is of quite recent origin. In the case of the pesticide chemicals, the insecticidal properties of certain plants had been known for some time, but identification of the active compound is of much more recent date, and it is only in the post-war years that they have been largely superseded by synthetic chemicals produced on a large scale by the chemical industry.

The natural radioactivity of uranium was first detected by the French scientist, Becquerel, in 1896, one year after Roentgen's discovery of X-rays. Several other radioelements, notably radium and polonium, were quickly discovered as a result of the pioneering work of Madame Curie. Until about 1945, two of these sources, X-rays and radium, both widely used in medicine, were the principal sources of human exposure over and above the natural background of radiation (see below, Fig. 4.1). This situation was to change dramatically after the discovery of nuclear fission in 1939, followed by the frantic development of the first atomic weapons during the war; the post-war years saw the development of atomic weapons of increasing size, and of nuclear reactors for power generation or for research. The radioactivity of all the fission products produced in nuclear fission, either in the testing of atomic weapons or in the operation of the power stations, greatly exceeds the total activity in use before the war, which was mainly in the form of the radio-element radium.

Fig. 4.1 illustrates the principal sources of human exposure to ionising radiations. The right-hand half of the figure shows all the natural sources of exposure for which quantitative data are presented in Table

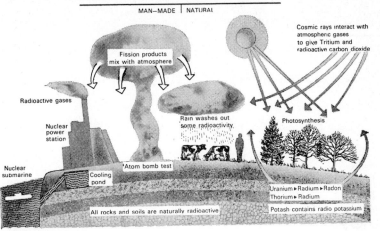

Fig. 4.1 Environmental radiation exposure: an illustration of natural and artificial sources of exposure

4.7 below. In the first half of the 20th century, the natural sources were augmented to a small degree by the fairly general use of X-rays and radium in medical practice.

The left-hand half of the figure shows the additional artificial sources of exposure which have resulted from the applications of nuclear fission in the development of atomic weapons and nuclear power reactors in the second half of the century. There are a number of much smaller contributions resulting from applications of a much greater range of radioisotopes in medicine, industry, research and teaching laboratories.

The quantities of radioactive material and pesticide chemicals have both multiplied rapidly in the post-war years. They have a further association in that their entry into the environment, whether deliberate or accidental, may result in their movement into food chains and their eventual entry into man through the food consumed. Both forms of pollutant may also interact with the biosphere in other ways and be a potential source of direct damage to the environment.

Apart from their arrival on the environmental scene together and their presence in food and in man, there are however, some very important and significant differences in the nature of these contaminants and their resultant effects. In the case of the pesticide chemical, the chemical itself, or an active metabolite, is the toxic agent and may, in the course of time, undergo certain biotransformations into another chemical form which is either harmless or rapidly eliminated from the body. In the case of a radioactive substance, however, the chemical substance itself may be completely harmless and its damaging effect is entirely associated with the radiations which it emits during its residence time in the body. Its radioactivity is quite unaffected by the state of its chemical combination or by any biotransformations which the substance may undergo. These may influence the site at which it is deposited in the body and its rate of elimination, but will in no way affect its activity as a radioactive substance. Radioactivity is a process of decay and the emission of the radiation is accompanied by the physical transformation of the parent radioelement into a daughter product which may be either a stable, non-radioactive element, or it may be another radioelement. The activity of any radioactive source will, therefore, decrease by physical decay in the course of time, each radioisotope having a characteristic halving time or half life. Any biotransformations then which affect a radioactive substance cannot change its activity, although they may result in a different distribution of the element within the body.

A further distinction between the two sources of pollution is that significant radioactivity of a radioelement is generally associated with extremely minute amounts by weight of the element, which would be quite undetectable by the most sensitive methods of chemical analysis available. An activity of 60 disintegrations per minute (dpm) of strontium-90 for instance, which represented the amount of strontium-90 ingested by the population of the UK daily in 1964, would only have weighed one-fifth of a micro-micro-gram, or one-fifth of a picogram, which may be compared with the daily intake of DDT amounting to 40 μg. This again serves to emphasise in the case of radioactivity that the chemical substance as such is not the toxic agent and that its harmful effects result solely from the radiations which it emits.

There is one further distinction to be made between the effects of pesticide chemicals and radioactive substances. The pesticide chemicals can only interact and affect the human body internally after they have been ingested from food and distributed within the various tissues of the body. Their effect is entirely limited to their presence within the body tissues. Radioactive substances, on the other hand, can prove harmful to man in two ways; firstly, as in the case of pesticide chemicals when the material has been absorbed and distributed in the body tissues. Secondly, the radiations emitted by the substance from its presence in the external environment can also interact with body tissues. This is due to the penetrating nature of some of the radiations emitted by certain radioactive materials. The two modes of radiation exposure are distinguished as internal in the former case and external in the latter case, and both types of exposure as in the case of radioactive fall-out may contribute to the total human dose of radiation. This chapter is concerned primarily with the effects of food pollution and the resultant internal dose to tissues.

RADIOACTIVITY AND IONISING RADIATIONS

PENETRATION IS just one of the qualities possessed by some of the radiations referred to in the previous section. Radiations representing various forms of radiant energy are universal and include the familiar light rays, the less familiar and invisible ultra-violet and infra-red radiations, micro-waves and radio-waves. These radiations are all part of the group of radiations included in the electro-magnetic spectrum, only a small portion of which is occupied by visible light. All these radiations, with the exception of some of the micro- and radio-waves, have low penetrations and are absorbed superficially by matter. They are normally regarded as harmless, although they can be extremely hazardous at high intensities. Laser beams, for instance, of visible light, ultra-violet, and infra-red radiations are exceedingly dangerous to the eyes, and excessive exposure to the sun's ultra-violet rays can also cause skin cancers.

The ionising radiations with which this chapter is concerned also have a range of qualities. They include the X-rays which are familiar from their well-established use in medical practice, and are also a part of the electro-magnetic spectrum, having good penetrating qualities. X-rays are frequently emitted by high voltage equipment, including television sets, especially the modern colour sets, but the radiation dose to viewers is extremely small (Fig. 4.10). This chapter is not, however, concerned with radiations emitted by special equipment such as X-ray sets, as these are solely a source of external exposure and can in no way produce problems of food pollution. The other ionising radiations include the highly penetrating γ-rays which are similar to X-rays, and less penetrating α- and β-rays. All three types of radiation, α, β and γ, are various forms of energy emitted during the spontaneous decay of radioelements. α-rays are typical of the heaviest elements, such as natural uranium and thorium, and the artificial element plutonium. β-radiation is emitted by a great majority of the artificial radioisotopes of elements, including the fission products. It is also emitted by some naturally occurring radioisotopes such as potassium-40 and carbon-14. The emission of γ-rays generally accompanies the α- or β-radiations. The essential differences between these three types of radiation are illustrated in Table 4.1. They are all invisible radiations producing no immediate sensory response so that there is no warning of exposure. They share the property of visible light and ultra violet rays however of fogging a sensitive photographic emulsion, a quality which is extremely valuable for their detection and measurement.

Table 4.1

Typical ranges of ionising radiations in air and water

Radiation	Range in air	Range in water
a	5 cm	30 μm
β	2 m	1 mm
γ	1 km	250 cm

Radioactivity, then, is a physical phenomenon, which involves the sudden release of energy which is trapped within the very tiny nucleus of an atom. All elementary substances are composed of atoms and all the atoms of a particular element are indistinguishable as far as their chemical behaviour is concerned. The atoms of an element may, however, differ in certain physical characteristics and when this difference is one of mass, the different masses represent the isotopes of the element. Isotopes therefore are atoms of the same element but distinguished by differences in their atomic masses. As an example, the natural element potassium (K) has three natural isotopes having mass numbers 39, 40, and 41, of which number 40 is a naturally occurring radioactive isotope. All atoms are characterised by a nuclear structure which may be likened to the solar universe with central sun and distant orbiting planets; they consist of a dimensionally very small nucleus surrounded by a cloud of negatively charged electrons (negative charges), at some distance from the nucleus and distributed at discrete energy levels. The chemical properties of the elements are related to the number and distribution of these electrons, all isotopes of an element having the same number and distribution.

Radioactivity, on the other hand, is a property of energetically unstable nuclei from which the surplus energy is emitted in the form of one or more of the characteristic radiations a, β or γ. The nuclei of radioisotopes can be identified by the quality of the ionising radiations emitted and the rate at which the disintegrations occur in addition to the difference in mass. When the disintegrations occur rapidly the half-life and lifetime of the radioisotope are short. The energy which is trapped within the particles which go to make up an atomic nucleus is far greater than the energy which is associated with the orbiting electrons. A great deal more energy is therefore released in the nuclear disintegration of an atom than is available from any chemical change involving its orbital electrons. The energy release on the atomic scale is even greater still whenever nuclear fission occurs.

Radioisotopes are then chemically identical with the stable element; apart from the difference in mass number they are only distinguished by the nature of their radioactivity. This identity of chemical properties will also be matched by an identity of biological behaviour, which will depend to a large extent on the chemical form of the element. Strontium-90 (Sr90) for example is a very important fission product which has no natural occurrence in the earth's crust, but is none the less chemically indistinguishable from the stable naturally occurring forms of this element which are all lighter in mass. Once the Sr90 has become associated with the natural strontium which occurs in very small amounts in the environment the two are no longer distinguishable chemically. In the case of a fission product such as Sr90 when the stable natural element is of low abundance, the movement of the radioisotope through the biosphere may also be influenced by other elements which have chemical affinities with the strontium. Such an element is calcium which can influence quite markedly the passage of Sr90 into food chains. Caesium-137 (Cs137) and natural potassium are a similar pair of fission product and related element.

This identity of chemical properties between a radioisotope and its stable element is also invaluable in assisting the recovery of the former for analysis and measurement. Although the activity of the radioisotope may be quite significant its actual amount in mass terms may be totally insignificant and not detectable by any conventional means. If, however, a known quantity of the stable element is added to a sample, and carefully homogenised with the radioisotope, the recovery of the latter is facilitated by normal chemical means. The radioisotope can then be determined by measuring its activity by one of the procedures described in the next section.

It should be noted that the majority of the chemical elements are completely stable consisting only of stable isotopes. A small number of elements are just the opposite consisting only of radioisotopes and without any stable isotopes. These elements which are generally found among the heaviest elements and include uranium, thorium, and radium, are referred to henceforth as radioelements. A number of natural elements are essentially stable elements consisting predominantly of stable isotopes but may include a small percentage of a radioisotope in their composition. They include elements such as carbon with one in every 10^{12} atoms being the radioisotope C14 and potassium with one in about every 10,000 atoms being the radioisotope K40. The fission products responsible for most of the artificial radioactivity found in food are mostly radioisotopes of stable elements.

DETECTION AND MEASUREMENT

IN VIEW of the lack of any awareness to radiation exposure, reliance must be placed totally on a range of physical devices for detecting and measuring the radiations. Fortunately, a variety of sensitive devices are readily available, ranging from sensitive photographic type emulsions to gas counters such as the Geiger-Muller counter and scintillation counting systems. The sensitive emulsions, which have been developed for X-ray work, after they have been exposed and developed are darkened in proportion to the amount of radiation exposure. Small dental X-ray films in a special badge holder are used especially for monitoring personnel who may be exposed as a result of their work. The Geiger-Muller counter produces an electrical signal which with the appropriate electronic units can be used to register individual radiations entering the detector. This counter is therefore a very sensitive detector for ionising radiations.

The scintillation counter employs an old device familiar from luminising paint in which a radioactive substance, for example radium, is mixed with a suitable phosphorescent chemical, for example zinc sulphide. The interactions of the radiations with the phosphor produce tiny scintillations of light which when scanned by a photomultiplier tube can be measured again as individual radiation events.

The measurement of radioactivity in food in the UK has been mainly the responsibility of the Agricultural Research Council, Letcombe Laboratory, who have regularly tested samples of milk and occasionally samples of cereals and animal produce.[1] The Fisheries Radiobiological Laboratory at Lowestoft has supplemented the work of the Letcombe Laboratory in the case of samples of freshwater fish, marine fish and edible seaweeds, especially in the vicinity of major nuclear establishments.[2]

RADIOACTIVE CONTAMINANTS IN FOOD

THE FOODS consumed by man have always contained small amounts of some of the radioactive elements which occur naturally in the environment and one of the more important of these is undoubtedly potassium. During the last thirty years, however, this natural contamination has been added to by a number of artificial radioisotopes. The more important of all the potential contaminants in food are listed in Table 4.2.

(a) Natural radioelements

THESE CAN be divided into three classes related to their origins. Firstly there are a few primordial radioelements which have been present in the earth's crust since its formation and have persisted ever since owing to their slow rates of decay and very long lifetimes greater than the age of the earth at about 4·5 billion years. They include elements such as uranium, thorium, and potassium. Secondly there are a number of radioelements usually present in very small amounts which owe their existence in the environment solely to their constant formation as daughter products of the long-lived primordial radioelements. A good

Table 4.2
Some radioactive contaminants of food

Class	Element	Mass no.	Radioisotope Symbol	Half-life	Origins
Natural	Carbon	14	C14	5760y	Cosmic ray activation product
	Potassium	40	K40	1300My	Primordial
	Rubidium	87	Rb87	47,000My	,,
	Polonium	210	Po210	138d	Daughter product primordial U
	Radium	226	Ra226	1620y	,, ,, ,,
	Thorium	232	Th232	14,100My	Primordial
	Uranium	238	U238	4500My	,,
Artificial	Carbon	14	C14	5760y	Activation product – A.W. + N.P.
	Zinc	65	Zn65	245d	,, ,, ,,
	Strontium	90	Sr90	28y	Fission product
	Ruthenium	106	Ru106	1y	,, ,, ,,
	Iodine	131	I131	8d	,, ,, ,,
	Caesium	137	Cs137	30y	,, ,, ,,
	Plutonium	239	Pu239	24,400y	Activation product

y = years; d = days; M = Mega, i.e. million
A.W. = atomic weapon; N.P. = Nuclear Power

example of this type of radioelement is radium-226 which is a daughter product of the long lived parent uranium-238. Radium has far too short a lifetime, about 1600 years, to have persisted since the earth was formed, but it is constantly renewed in the earth's crust by the decay of uranium.

The radioactive gas radon is a further daughter product of radium; it is forming constantly from radium present naturally in rocks and soils, and as it is a gaseous element it emanates from the ground into the atmosphere. There are always measurable amounts of this gas in the atmosphere and its concentration is augmented whenever coal is burnt and during times of temperature inversion. The radioactive decay of the radon eventually leads to the formation of two rather more persistent and radioactive daughter products, lead-210 and polonium-210. These two elements form solid compounds and are consequently washed out of the atmosphere by rain descending to the ground as a natural radioactive fall-out.

Radioisotopes which are constantly forming in the atmosphere by the interactions of cosmic radiation form the third type of naturally occurring radioisotope. Cosmic rays are also a form of ionising radiations of extremely high energy. They enter the earth's atmosphere from outer space, and carry so much energy that they interact quite violently with the nuclei of atoms normally present in the atmosphere. These interactions transmute the target element into a radioactive product, and one very important reaction of this type is the transmutation of atmospheric nitrogen to form radioactive carbon-14, $C14$. This mixes with carbon dioxide in the atmosphere and enters a food chain by photosynthesis in plants.

All of these natural radioelements exist in minerals, rocks, in soils, and in surface waters, although the atmosphere is the major source of the $C14$. They can enter into a food chain from the soil, water, or atmosphere and may therefore be present in the food consumed and hence in the human body as natural sources of radioactivity (Table 4.3).

Table 4.3

Human consumptions of some natural radioisotopes

Radio-isotope	Average intake pCi/day	Average amounts in body, pCi
Ra 226	1–2	50
Pb 210	1–10	250
K 40	2800	125,000

Natural radioactivity in the body is responsible for about one-quarter of the total background dose of radiation to which the body is normally exposed from all natural sources. This represents the internal fraction of the total dose, the remaining threequarters being an external dose contributed partly by the cosmic radiation and partly by the radiations emitted by the natural radioelements surrounding us in the rocks, soils, and building materials. Potassium-40 is responsible for the major part of the internal dose, with a much smaller contribution coming from the carbon-14. A fraction of the radium which is ingested in food is deposited in the mineral bone and may contribute a further small localised dose to the bone and the bone marrow (Table 4.7).

(b) **Artificial radioisotopes entering into food**

ARTIFICIAL RADIOACTIVITY in the environment has increased in the post-war period. The larger quantities of artificial radioactivity distributed on a global scale have resulted from the testing of atomic weapons and especially the several series of massive testing programmes in the early 1960s. Much smaller quantities of artificial radioactivity enter the local environment continuously as wastes discharged from major nuclear installations such as the nuclear power stations and the chemical industries which support the power programme by fabricating the fuel elements and processing the irradiated and highly radioactive fuel elements after their operational period in the reactor core. An even smaller contribution arises from the low level radioactive wastes discharged from hospitals and research laboratories. These origins of the more important radioactive materials entering into the environment are represented diagrammatically in Fig. 4.2, and the location of the major nuclear installations in the UK is shown in Fig. 4.3.

The number of radioactive substances entering the environment from these various sources is considerable and they are produced in two distinct ways. Both atomic weapons and nuclear reactors produce a complex mixture of radioactive waste products which include the fission products and various activation products which are produced during the fission process. The various radioisotopes for medical, industrial and research purposes are also produced by activation processes but using carefully selected and controlled nuclear reactions. It is convenient to consider the various artificial environmental contaminants in the separate categories of fission products and activation products.

Nuclear fission is a property possessed by certain heavy atomic nuclei such as uranium-235 or plutonium-239. Natural uranium consists mainly of the heavier isotope U238 and only about 0.7% of the lighter U235. Plutonium is an entirely artificial radioelement, the isotope Pu239 being formed by nuclear activation reactions whenever the fissioning of U235 is taking place in natural uranium. Nuclear fission is initiated by the capture of a nuclear particle (a neutron) and involves the splitting of the heavy nucleus into two asymmetric fission fragments. These two

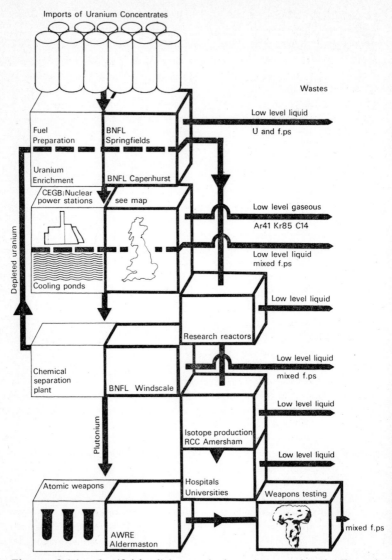

Fig. 4.2 Origins of artificial radioisotopes in the environment (see also Fig. 4.3)

fragments are the radioactive isotopes of lighter elements, and when a mass of fissile material is undergoing fission, the fissions can take place in a variety of ways producing a highly complex mixture of fission products. The process of fission is also accompanied by the prompt emission of penetrating radiations and the release of a large amount of energy which is very largely converted into heat. In addition to the fission products the intense gamma radiation and energy release, each individual fission event also produces two or three fresh neutrons and provided that at least one of these neutrons is available to induce a fresh fission a self-sustaining chain reaction will be set up. When the chain reaction spreads with great rapidity through a critical mass of the pure fissile material due to a large number of simultaneous fissions there is an explosive release of this energy and an atomic detonation is the result, a whole mass of fission products being produced in the simultaneous burst of fissions. When the chain reaction is controlled in the core of a nuclear reactor the energy release is also regulated and the fission product wastes can be contained in the metal-clad fuel elements. The energy, as heat, can be extracted from the reactor core by circulating a coolant gas or liquid and the hot coolant used to raise steam, drive the steam turbines and generate electricity.

The activation products by contrast are not produced directly in the fission process although they are formed at the same time by the interaction of radiations (mainly neutrons) produced in fission with stable elements which may be drawn into the atomic burst or which may be part of the structure of a weapon or of a reactor core. Plutonium-239 is produced in this way from the heavier isotope of uranium whenever natural uranium is used in the reactor fuel elements. The activation products may also include a wide range of radioisotopes some examples of which are listed in Table 4.2.

The fission products and the activation products which are formed in large amounts in the nuclear power programme, are far too complex in composition to make them suitable as sources of radiation or of radioisotopes which can be used in medicine, industry, or research. They constitute a highly radioactive form of waste, far too active for disposal to the environment, so that arrangements have to be made for their safe storage for many years. When radioisotopes are therefore required for special applications in medicine, industry or research they have to be made by activation processes with selected target materials irradiated under controlled conditions. These conditions can be arranged close to the core of a research or specially designed reactor. In this way it is possible to activate a great majority of the natural stable elements converting them into radioisotopic forms, suitable for a variety of special applications. In the UK, these activations are generally carried out in reactors at the Atomic Energy Research Establishment at Harwell for processing and marketing by the Radiochemical Centre Ltd at Amersham. The centre is currently marketing about 400,000 consignments of radioactive materials consisting of a large number of radioisotopes and labelled compounds with a wide range of activities.

Intensive irradiation treatments of a very different type are also

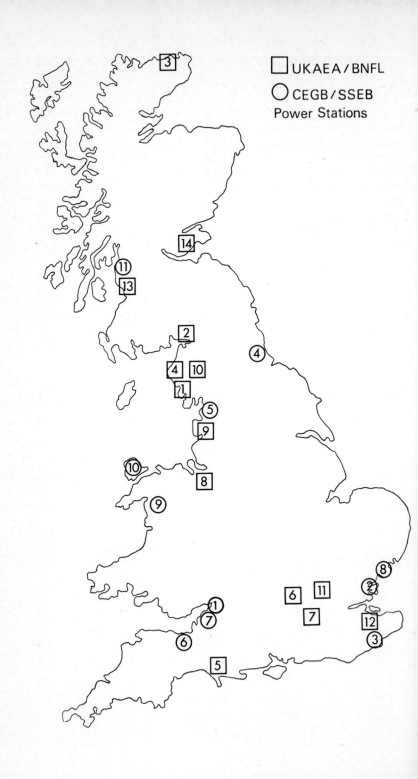

□ UKAEA / BNFL
○ CEGB / SSEB
Power Stations

applied to the sterilisation of medical instruments and pharmaceutical packages, and have also been approved in a few countries for the bacterial sterilisation of selected foods in order to increase their shelf life. These foods include some meat products such as bacon and also potatoes. The irradiations are carried out in specially designed plants containing massive amounts of the activation product cobalt-60, producing intense γ-radiation fields. Although these foods may be subjected to extremely intense radiation, they are not activated by the γ-rays, which do not have sufficient energy to induce nuclear reactions leading to radioactive products in the food. Although the process is therefore safe in producing no radioactive contamination in the product, it can create other problems affecting the quality of the food, such as its texture, appearance, flavour, nutritional value and even more importantly it may introduce traces of toxic chemicals resulting from radiation-induced chemical changes in the food. This latter possibility is a matter of some concern calling for long-term investigations before the irradiated product can be approved as safe for human consumption and is the main reason why so few food irradiations have been officially approved. The process has the advantage that it can be used after the food has been packaged.

RADIOACTIVE WASTES FROM ATOMIC POWER STATIONS

THE DEMAND for energy increases at a faster rate than the growth rate of the population. The doubling time for generating capacity in Britain appears to be about 15 years, with a total capacity in 1972 of 55,000MW and about 10% of this in atomic power stations. The

---

Fig. 4.3 Location of major nuclear installations in the UK

CEGB Power Stations:
1. Berkeley   Magnox
2. Bradwell   Magnox
3. Dungeness A   Magnox
   Dungeness B   AGR
4. Hartlepool   AGR
5. Heysham   AGR
6. Hinkley Pt. A   Magnox
   Hinkley Pt. B   AGR
7. Oldbury   Magnox
8. Sizewell   Magnox
9. Trawsfynydd   Magnox
10. Wylfa   Magnox

SSEB Power Station:
11. Hunterston   Magnox

UKAEA Power Stations:
1. Calder Hall   Magnox
2. Chapelcross   Magnox
3. Dounreay   PFR
4. Windscale   AGR
5. Winfrith   SGHWR
                HTGR

UKAEA Research Reactors:
6. AERE Harwell
7. AWRE Aldermaston

Brit. Nuc. Fuels Ltd. (BNFL):
8. Capenhurst Works
9. Springfields Works
10. Windscale Works

Radiochemical Centre Ltd.:
11. RCC Amersham

Ministry of Defence (Navy Dept.):
12. Chatham, HM Dockyard
13. Faslane, HMS Neptune
14. Rosyth, HM Dockyard

forecasting of future energy demands is a very difficult exercise, but it has been estimated that a total of about 100,000MW, with approximately one-quarter as nuclear capacity, will be required in 1980. There seems to be little doubt that in Britain as elsewhere there must be increasing reliance on and a fairly rapid expansion of nuclear capacity in the near future.

Nuclear power reactors can be designed in many ways, but in Britain the preference in the early phase of the programme has been for reactors which are gas cooled, graphite moderated and with natural uranium fuel elements clad in magnox alloy. All the nuclear generating capacity operating at present is based on this system, with the next planned stage of development being represented by the Advanced Gas Reactors (AGRs). These reactors are similar to the first generation of reactors, but slightly enriched uranium is substituted for natural uranium, and stainless steel cladding replaces the magnox cans in view of the higher core temperatures which are reached. A number of reactor stations of this type are currently under construction, but there is much uncertainty about the next stage of development, which will be required to bridge a gap before the fast breeder reactor systems are fully available. The British Government announced in the summer of 1974 its decision to build a number of Steam Generating Heavy Water Reactors (SGHWR) as the next immediate phase of the Nuclear Power Programme. An experimental version of this reactor is operating at Windfrith, and the system has similarities with the Canadian Candu Reactors. The fast reactor is expected to play a major role in future plans for energy generation, and a prototype is currently under construction in Britain, to operate on fully enriched uranium-235 or plutonium-239 fuels. The fast reactor system employs a molten metal, sodium, as the primary coolant to extract the intense heat from the core. A blanket of non-fissile uranium-238 or thorium-232 is arranged around the core and is converted during the reactor operation into fresh fissile fuel – plutonium-239 or uranium-233 respectively.

Various alternative systems of nuclear power reactors have been designed and in the USA especially light water reactors (LWR) of two types, have been the basis of the nuclear power programme. In both cases ordinary water serves both as the coolant and the moderator. In the case of the boiling water reactor (BWR) the steam generated is used directly to drive the turbines, and in the pressurised water reactor (PWR) the hot water first flows through a heat exchanger; the steam from the latter being fed to the turbines. In both cases slightly enriched uranium is used as the fuel. As in the case of the British Magnox reactors, considerable operating experience has been acquired with the light water reactors and they have now been adopted by several other countries. In the USA there has, however, been considerable discussion about their safety and especially the arrangements for the emergency cooling of the reactor core in a loss of coolant accident.

Ultimately it is hoped that reactors based on nuclear fusion will be developed to the stage where they can be used for controlled power generation. The fusion reaction which is also the basis of the hydrogen atomic bomb and the source of all solar energy, has one great advantage over power systems based on nuclear fission, in that only small quantities of highly radioactive wastes will be produced. These fission product wastes are a major problem in the successful development of nuclear power based on fission, although the problem of their recovery and storage is being tackled in a variety of ways in several countries.

The fuel elements for the British power reactors consisting of either natural or slightly enriched uranium, are fabricated at the Springfields Works of British Nuclear Fuels Ltd (BNFL). The elements are clad in metal, either Magnox Alloy in the first generation reactors, or stainless

Table 4.4

Some discharges of liquid radioactive wastes to surface and coastal waters, based on Technical Reports of the Fisheries Radiobiological Laboratory, Lowestoft[2]

Establishment	Site	Annual total activity of discharges	
		Authorised Ci	Actually discharged in 1970 Ci
CEGB nuclear power stations	Berkeley*	200	24
	Bradwell*	200	128
	Dungeness*	200	84
	Hinkley Point*	126	126
	Oldbury*	100	8
	Sizewell*	200	24
	Trawsfynnydd*	40	14
	Wylfa*	65	7
SSEB nuclear power stations	Hunterston*	200	64
BNFL	Springfields	12360	980
	Windscale†	306000	123000
UKAEA	Dounreay†	24240	15240
	Winfrith†	31200	1200
Ministry of Defence	Chatham*	20	0·2
	Rosyth*	20	0·3

* Excludes an allowance for tritium
† Includes an allowance for γ-emitters

steel in the case of the AGR, in order to contain all the fission products which are formed during the operation of the reactor. After several years of operation they are then discharged and replaced by fresh fuel elements to maintain the efficiency of the reactor. The fuel elements which are discharged from the reactor contain an accumulation of highly radioactive fission products (about 500 MCi at present power levels) and have to be stored temporarily in cooling ponds at the power stations to allow a major part of the radioactivity due to shorter lived fission products to decay. Small quantities of longer lived fission products may escape into the cooling waters during this time and eventually be discharged into the environment. After the cooling off period the fuel is transferred to the BNFL Chemical Reprocessing Plant at Windscale Works. The uranium and plutonium are recovered and the accumulated fission product waste is concentrated for storage in special double-lined water-cooled tanks. Much smaller quantities of low level wastes arising on the plant which it would be extremely expensive to remove, are finally discharged via a pipeline into the north Irish Sea. All the discharges which occur regularly at the power stations and at Windscale Works are controlled so as to minimise the most serious risk of food contamination which can be identified in the area of the discharge, and they are also regularly monitored (Table 4.4).

The advent of light water reactors might also increase the variety of radioisotopes in the regular discharges of gaseous wastes normally emitted at all the power stations and at Windscale Works. In the case of the gas-cooled reactors this is mainly argon-41 which is an activation product formed from the natural element in the air used to cool the outside of the pressure containment vessel. Quantities of krypton-85 and some xenon isotopes are also released at Windscale. Some regular releases of krypton-85, iodine isotopes and possibly caesium-137 may also occur regularly from the water reactors.

The uranium which is recovered at Windscale Works is depleted in the fissile isotope, U235. This uranium and fresh uranium from Springfields Works can be processed at the Capenhurst Works of BNFL to produce uranium in which the original isotopic composition is restored or which may be further enriched in U235 for the AGR programme. The process produces a quantity of uranium which is considerably depleted in U235 and is of no further value for the power programme. This depleted uranium may find some applications as a shielding material. The process at Capenhurst Works employs gaseous diffusion to achieve the enrichment of the uranium, but a gas centrifuge is under development to provide a cheaper and more efficient process.

The quantities of the principal wastes arising from the whole of the British Nuclear Power programme and authorised discharges are summarised in Table 4.4. These various forms of radioactive waste from the power programme receive a very small addition from the growing practice of using radioisotopes in hospitals for medical diagnosis or therapy, in industry for investigating process problems, and in teaching and research laboratories where they have many useful applications. All these practices use a considerable number of artificially produced radioisotopes, supplied by the Radiochemical Centre Ltd, of Amersham, but generally in very minute amounts compared with the levels of waste discharged from the nuclear establishments involved in the power programme. The discharges of radioactive wastes from hospitals, industry and laboratories either in the form of liquid wastes into sewers or in the form of solid wastes to be buried on local authority tips, which are all permitted and authorised under the Radioactive Substances Act of 1960, are most unlikely to lead to any problems of food contamination. They are also quite insignificant compared with the total quantities of highly radioactive fissio products released into the atmosphere from all the atomic weapons which have been tested above the ground to date.

ATOMIC BOMB FALL-OUT

THE 1950s were a period of intensive weapons testing, on land, sea, and in the air culminating in two major series of tests during 1961 and 1962. These tests aroused considerable controversy among scientists regarding the long-term effects of the consequent radioactive fall-out and also aroused growing anxiety among the general public. Many measurements of fall-out and investigations into its transfer through the biosphere into man were initiated during this period; they gradually provided invaluable information about transfer mechanisms, the activity levels in the principal foods and the possible consequences for man. The problems of radioactive contamination have as a consequence received a great deal more attention than the other forms of food contamination; the studies have provided valuable guide lines however for the investigations into these other contaminants. A further consequence of the work on radioactive fall-out and improvements in the techniques of investigation is the stimulus it has provided for studying the behaviour of natural radioactivity in the biosphere. Radioactive fall-out shares with DDT and other chlorinated hydrocarbons the doubtful distinction of artificial pollution on a global scale.

The warnings of certain scientists, and notably Linus Pauling, with backing from public opinion were instrumental in securing a test ban treaty, with a moratorium on all testing of atomic weapons at the surface of the earth or in the atmosphere as from the end of 1962. The treaty was signed by the three major nuclear powers at that time, and has only been broken sporadically by two other countries anxious to join the select group of nuclear powers. These occasional tests, augmented by some leakages from underground tests, have continued to add further small amounts of the fission products to the considerable reservoirs in the environment from the earlier massive series of tests. Two of these fission products, Sr90 and Cs137 are prominent in the pools of artificial radioactivity owing to their relatively long radioactive half-lives. This chapter is concerned mainly with a detailed analysis of their behaviour in the environment, the present and future levels in food and in man, and the possible consequences of these levels for the future health of mankind. Other fission and activation products which pose different types of problem either of shorter duration, e.g. iodine-131 (I131), or of a more localised character, e.g. ruthenium-106 (Ru106), are also referred to.

The patterns and rates of deposition of fission products at the earth's surface after a nuclear explosion depend on various factors such as the latitude of the test site, the height above the ground, the size of the weapon, and prevailing weather conditions. In the case of the original kiloton explosions the fission product cloud is injected only into the lower atmosphere, the troposphere; the contaminated cloud circulates around the earth in an easterly direction broadening about the line of latitude, with deposition being virtually complete in about two months. The major part of all the fission products released up to 1962 were however injected into the upper atmosphere, the stratosphere, from which the deposition rate at the earth's surface is much slower. The high yield megaton tests i.e. the hydrogen fusion bomb initiated by a fission bomb core are responsible for injections into the stratosphere; the mixture of fission products largely in the form of vapour, condenses into an aerosol cloud of very fine particles having sub-micron dimensions, or is absorbed on to dust particles drawn into the mushroom cloud. This aerosol cloud settles only slowly and is transported largely by air movements, resulting in broad global distribution patterns. There is some settling under gravitation as some of the particles coalesce, but the mechanisms responsible for the eventual transfer of all the activity to the troposphere are quite complex. Once the particles have entered the lower atmosphere they are deposited fairly rapidly; air turbulence brings the particles down to the rain-bearing clouds, where the process of deposition is completed by rainwater. Some of the particles may serve as nuclei for the condensation of the raindrops; or they may be washed out by the falling rain. The bulk of the activity arriving at ground level is therefore brought down by rainwater; by contrast the dry deposition rate is very slow and only assumes importance in low rainfall regions. The rate of deposition at the earth's surface is largely controlled by the rate of transfer from stratosphere to troposphere with an average residence time in the stratosphere of about two years. Once a fall-out particle is in contact with rainwater, much of the fission product activity and most of that due to strontium-90 and caesium-137 will pass into the solution phase.

Maximum deposition of the two fission products Sr90 and Cs137 has taken place in the northern temperate zones, within a broad band between latitudes 30° and 70° north, resulting largely from the major series of tests in the northern hemisphere during 1961 and 1962. Within the broad pattern of fall-out there are more localised variations mainly associated with differences in the annual rainfall; high ground with westerly facing slopes is exposed to higher contamination levels for

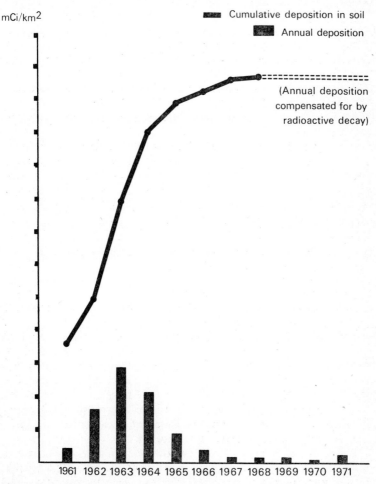

mCi/km²

▬▬▬ Cumulative deposition in soil

▨ Annual deposition

(Annual deposition compensated for by radioactive decay)

1961 1962 1963 1964 1965 1966 1967 1968 1969 1970 1971

Fig. 4.4 Deposition of Caesium-137 normalised to an average rainfall of 95cm³ (One division of the scale = 10 mli k/m²)

example in the UK. There are also seasonal variations with maximum fall-out tending to occur in the late spring or early summer months. The annual fall-out pattern for Cs137 which may be regarded as typical of the northern temperate zones is illustrated in Fig. 4.4. Estimates of Sr90 fall-out are obtained by dividing the Cs137 values by the factor 1.5. A peak in the annual deposition in 1963, followed by a rapid decline to 1967, with a subsequent levelling off should be noted. This levelling-off is due to sporadic testing by two countries after the suspension of testing by the major nuclear powers, and has arrested the decline which would have continued beyond 1967.

In addition to the long-range global fall-out there will always be rapid close-in or local deposition after an explosion, and this could be a particularly serious problem in any disastrous outbreak of nuclear warfare. This close-in fall-out would give rise to high external radiation levels for some days, and would create urgent problems of food contamination mainly from short-lived fission products, such as the iodine isotopes and especially I131. Subsequently the much longer-term problems from the more persistent strontium and caesium isotopes would claim major attention.

Similar problems might also arise in a civil emergency involving a nuclear power station, but would probably be even more localised in their effects. The probability of such an event is extremely low in view of the extraordinary safeguards built into the reactor, but this low probability is partly offset by the increase in the number of stations in Europe or a small country like the UK.

RADIOACTIVE CONTAMINATION OF AGRICULTURAL PRODUCE

(a) Strontium-90 and Caesium-137

THE PRODUCTS of agriculture provide a major part of the food consumed by man, with lesser contributions from the seas and freshwaters. Particular attention must be given therefore to agricultural produce when attempting to assess the average population dietary intakes of the two fission products Sr90 and Cs137. The contamination levels in fish and other aquatic produce may have to be considered for much smaller population groups who may be subsisting to a much greater extent than the average population on these foods. The critical pathways representing the movement of environmental fission products through the more important food chains are set out in Fig. 4.5.

Plants are the ultimate source of all the foods consumed by man and produced in agriculture. They are therefore an important first link for the entry of the contamination into the food chain and ultimately into man.

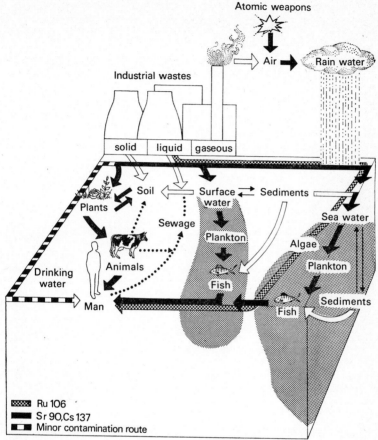

Fig. 4.5 Radioactive food chains

Fig. 4.6 illustrates the stages in agricultural production which result in produce contaminated by radioactive fall-out. These contaminants enter into all the principal foods but milk, dairy produce, meats and to a lesser degree, cereal grains, are the major sources in the human diet. Their presence in milk is of special concern for young children. The iodine isotopes present a short-term problem of about one month's duration immediately after a release of a fresh mixture of fission products from an atomic weapon detonation or a possible accident at a nuclear installation. The much longer-lived Strontium-90 and Caesium-137 present the longer-term problems.

Radioactive contamination of plants growing out of doors can occur by two distinctive routes:

Fall-out Routes to Man

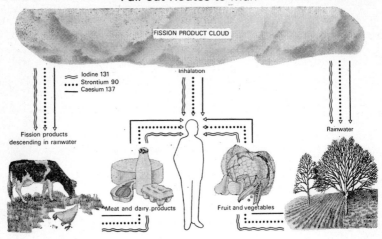

Fig. 4.6 Fall-out routes to man: an illustration of the stages in agricultural production which result in produce contaminated by radioactive fall-out

1. Directly by the action of contaminated rainwater falling on to exposed upper parts of plants.
2. Indirectly by processes of transfer from the accumulation of the contaminants in the soil via the root system of the plant.

The first route is always predominant at times of high fall-out soon after the testing of atomic weapons, and this is especially so for the short-lived iodine-131. In view of the short half-life of about eight days, most of this activity has decayed by the time that activity has been transferred from the soil, and the plant harvested. This second route, however, will increase in importance for the long-lived fission products as the time between testing and cropping is also increased. At a certain stage it may be the sole or major source of plant contamination.

The transfer of any accumulation of the contaminant from soil to plant will depend markedly on the characteristics of the soil and the rooting system of the plant. The soil characteristics are influenced by the composition of the soil and especially the content of clay and organic matter, and they may also be modified by the application of mineral fertilisers.

The soil characteristics influence mainly the availability of a particular fission product in the soil solution, whereas fertiliser treatments influence the availability of essential soil nutrients which may modify the uptake of the fission products.

Adsorption on to soil components, especially organic matter and clay mineral particles is very likely to occur and to be a vital factor in removing activity quite rapidly from the soil solution. Caesium is adsorbed very rapidly and held very tenaciously by clay minerals in most soils to become practically unavailable to plants through the normal processes of root absorption. Although the organic matter may also remove some of the caesium from solution there is evidence that rather more of the fission product may be made available in the presence of high organic contents. $Cs137$ is therefore generally immobile in a majority of soils and tends to accumulate in the top layers especially of undisturbed soils such as permanent pastures. Strontium is also partially removed from soil solutions but appears to be less tenaciously held by the clay particles than is the caesium. It is therefore rather more mobile within the soil layers, penetrating to greater depths of the soil profile, and only being partially removed from the water draining through the soil. It is also rather more readily available to plants through their roots than is caesium, and at times when fresh fall-out is low, this may be the most important source of contamination reaching the plant.

This transfer of caesium and to a lesser extent of strontium on to the surfaces of fine soil particles can lead to a fairly rapid reduction of their concentration in rainwater as it collects and flows into streams, reservoirs and other surface waters. Rainwater percolating through the soil and collecting in underground reservoirs also undergoes a considerable decontamination. As a consequence, drinking water concentrations of the two radioisotopes are generally extremely low and only a very minor source of human contamination as compared with food. As a precaution, however, they are regularly monitored by the Government Chemist.

Applications of mineral fertilisers and liming of the soil generally minimise the uptake of fission products as the soil solution will almost certainly contain higher concentrations of the two elements potassium and calcium with which caesium and strontium respectively have the closest chemical affinity. Whenever the two stable elements are present in the soil solution the root system will discriminate against the heavier radioisotopic element so that the ratios of caesium activity to potassium and strontium activity to calcium are normally reduced in the transfer process from soil to plant. This ability of the root system to discriminate in favour of the lighter stable element is much reduced in soils of poor quality such as those in permanent hill pastures which are often acidic and deficient in the supply of the essential elements. Relatively high uptakes of the two fission products and especially of strontium from the soil by the herbage may then result. Animals grazing on impoverished hill pastures which may be subject also to higher rainfall and fall-out may also be exposed to higher levels of contamination compared with animals on lowland pastures.

The direct contamination route has, however, been responsible for most of the contamination in plant produce to date. In this case the factors which influence the extent of plant contamination are much more dependent on the characteristics of the plant such as its size, shape and configuration of leaves, flowers and fruiting systems. The total surface area presented by the plant to rainfall is especially important as also is the extent of rainfall occurring at particular stages of the plant growth and more especially when the crop is ripening just before harvest. Any activity in the rain which is still in a particulate form may lodge on the plant surface and remain there until it is consumed by animals or man. Any activity which is in solution may adsorb on to the surface, and as the actual mass concentration of the fission products is also extremely low, the two isotopes are transferred rapidly and completely from the solution to the plant leaf surface. In this case there is no competition with stable elements and any caesium adsorbed in this way may then transfer fairly quickly through the surface membrane of the leaf to be translocated quite rapidly to most parts of the plant. Strontium, on the other hand, penetrates the membranes rather less readily and may remain mainly at the surface.

Climatic factors such as the prevalence of wind and sunshine, humidity and temperature, may also affect the extent of final contamination by this route. The major importance of direct contamination at times of fresh fall-out is illustrated by the following data for grass, about 1963–4, when fresh fall-out was high:

Sr90   80% by direct foliar contamination
         20% by indirect root contamination
Cs137   99% by direct foliar contamination
          1% by indirect root contamination

The more important plant sources of human contamination are the various cereal grains and their products such as flour. Samples of grain have been examined regularly by a number of laboratories, notably the Danish Atomic Energy Commission Research Laboratory at Riso[4] and the Letcombe Laboratory of the ARC[1] and supported by some analyses at the University of Manchester Radiological Protection Service.[5] A compilation of several series of results for Cs137 in cereal grains is presented in Fig. 4.7, and shows the rapid decline in fall-out levels of the produce from the peak values of 1963–4 and the subsequent levelling off after 1967. Small variations in the levels since 1967 may be attributed to the sporadic testing of weapons referred to earlier. The distribution

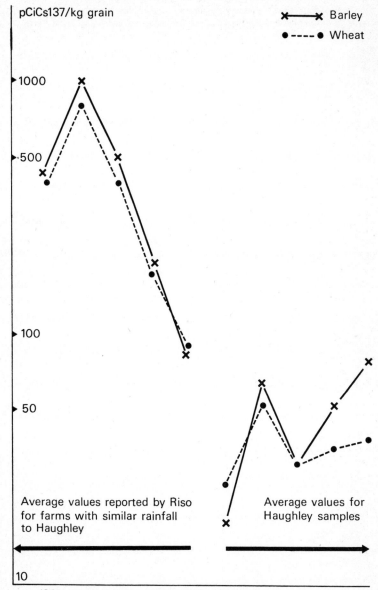

Fig. 4.7 Some values of Cs137 in cereal grain samples

of the activity within the various parts of the plants and in the whole grain is not uniform, the pattern depending on the amount of rainfall and the concentration of activity especially when the ears of grain are ripening in the field. During the major growth period of the plant the leaf area traps the greater amount of the activity, and the levels in the grain are only about one-quarter to one-third of the total contamination in the whole plant. At times of relatively high fall-out and wet autumns rather more activity will be found in the grain and most of this activity will be in the outer parts or husks. It has been shown, for example, that wheat flour of 70% extraction contains only about one-third of the caesium in the whole grain, and about one-tenth of the total strontium content of the whole grain. Wholemeal flours are second only to animal produce as a source of these two fission products in the human diet. Green leafy vegetables may also be important at times of fresh fall-out, but root vegetables and fruits are usually far less important.

All authorities are however agreed that animal produce, and especially fresh milk and milk products, are the major dietary sources of the long-lived fission products. For this reason milk and some of its products have provided the major part of any regular monitoring programme in many countries. In the U.K. the responsibility for milk analyses has been undertaken by the Letcombe Laboratory of the ARC and their main programme has been to assess the national averages of milk contamination. The activity levels in milk provide also a valuable guide to the total dietary contamination, enabling a considerable simplification in sampling and analysing contamination in the whole diet. The contamination in animal produce derives from the contamination in the plant food supply of the animal, which for farm animals is mainly grass and grain supplements. Whereas the contaminants in the rainwater are unaccompanied by any significant amounts of other stable elements, except possibly calcium, the situation within the plant is similar to that in soil. The active elements are in the presence of much higher concentrations of the elements potassium and calcium, which are essential mineral elements for both plant and animal. The active element is in competition with the stable elements during the digestion of the plant by the animal, and is to some extent discriminated against in passing into the animal tissues and in elimination via the kidneys. So far as is known, living organisms make use of the lighter elements in preference to the heavier elements, although there may be some exceptions as far as a few trace elements are concerned. The final levels in the animal are still sufficiently high however to be a really significant source of human contamination, as animal produce forms a major part of the average human protein consumption. Values for milk, based on the data published by the Letcombe Laboratory, are presented in Fig. 4.8.

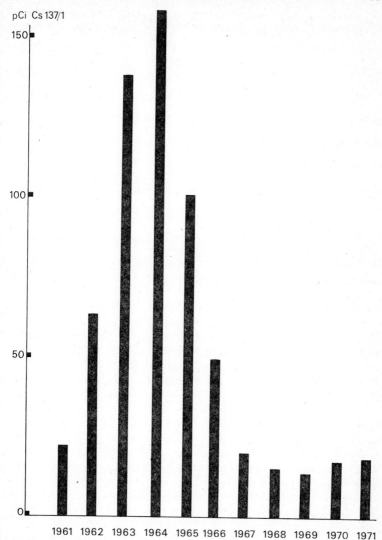

Fig. 4.8 Country-wide average values for Cs137 in milk (results reported by the Letcombe Laboratory, 1971 Annual Report[1])

Discrimination factors which relate the concentration of a fission product in the animal produce to its concentration in the animal feeds have been established in a number of cases. They are generally expressed as an Observed Ratio (OR), of which the following are examples:

$$\text{OR for Sr90 (milk/feeds)} \quad = \quad \frac{\text{SU in milk}}{\text{SU in feeds}} = \quad 0\cdot11$$

$$\text{OR for Cs137 (meat/feeds)} \quad = \quad \frac{\text{CU in meat}}{\text{CU in feeds}} = \quad 1\cdot6$$

(See the end of this chapter for definitions of units)

(b) Other radioisotopes

THE BEHAVIOUR of strontium-90 and caesium-137 as the principal artificial radioactive contaminants of food at the present time has been reviewed in some detail. They are not, however, the only artificial radioactive contaminants in food; at other times and in other circumstances other radioisotopes may assume greater significance.

The fission product iodine-131 may be a significant hazard in the early stages of any event leading to a release of fission product activity and it may subsequently follow a similar critical pathway to that of caesium and strontium through the environment. It is produced in quite high yield during the fission process and may therefore be released in quite large amounts, especially in local fall-out in the early weeks after a test or any major nuclear emergency involving a nuclear reactor or processing plant. It is also a volatile element which may facilitate its wider dispersal after a release. It may then reach fairly high levels in milk and fresh milk products and may present for a limited time a serious hazard especially to young children, as iodine is an essential element concentrating in small amounts in the small thyroid gland. It is, however, a short-lived radioisotope with the result that any contribution to food contamination is likely to be much reduced about one month after the release. During this early period it could cause more serious problems than any caesium or strontium that may accompany it. The accident involving one of the original atomic reactors at Windscale Works in 1957 illustrates some of the problems that might arise during a large scale release of iodine-131.[6] Because of its volatility, it was the only fission produce to escape in significant amounts during the fire in the core of the reactor. Deposition of the iodine from close-in fall-out within a radius of about 20 miles of the Works was sufficiently high to lead to milk contamination in excess of recommended levels for safe human consumption so that milk deliveries within the area had to be suspended for about one month. There was no reason why the milk should not have been processed into cheese and butter however, as any activity in the dairy produce could have been allowed to decay before consumption.

Another much more persistent radioisotope which is formed in an atomic explosion as an activation product is carbon-14. As indicated in Table 4.2 this is also a naturally occurring radioisotope with about one in every 10^{12} atoms of natural carbon being in the radioisotopic form. The radioactive C14 content of the carbon dioxide in the lower atmosphere in the years 1963 and 1964 increased by about 90% over the natural level. This led to an estimated increase in the average UK diet of about 80% about twelve months later and an increase of about 70% in the body tissues a further year later. The lower increase in man and the delay in reaching the maximum increase is due to the gradual dilution of atmospheric radioactive carbon dioxide by the much greater mass of carbon in the biosphere and also to gradual losses to the oceans. All the concentrations have decreased since the maximum values as a result of these factors, and also the combustion of fossil fuels releasing nonradioactive carbon dioxide into the lower atmosphere. This increase in the activity of the vital element, carbon, does not contribute a great deal extra to the total human radiation dose, but as it is likely to be incorporated in all or any of the organic substances which are part of the tissue cells, its presence may have a greater significance than the actual radiation dose delivered to the tissues. It may for example be incorporated in the molecular structure of the DNA in the cell nucleus, and if it should decay as a part of the DNA, the resultant nuclear transformation would result in the formation of a nitrogen atom at the expense of the original C14 atom, and also the rupture of chemical bonds; a point mutation or chromosome aberration could be an almost inevitable consequence. This would be additional to damage which might result from the radiation emitted in the disintegration. Similar problems arise from tritium, the heavy radioactive isotope of hydrogen, which is also produced in weapons tests and in the nuclear reactor power programme.

Some isotopes of the purely artificial element plutonium have also been released into the environment in recent years. Plutonium-239 is an activation product of the heavier isotope of natural uranium in the fuel elements and is produced in quite large amounts during the fuel cycle within the reactor. It is a fissile isotope and it can be used to construct an atomic weapon or as an alternative to uranium-235 as a nuclear fuel, especially in the future fast reactor programme. It is therefore an important product of the chemical reprocessing of the irradiated fuel elements, from the power programme.

Pu239 was used as the nuclear explosive in many of the atomic weapons tests including one of the original explosions over Japan. (The alternative nuclear explosive is of course U235 which has to be separated from natural uranium and involves an expensive operation on account of its small content, 0·77%.) Owing to the very rapid and tremendous release of energy in the explosion, some of the Pu239 is dispersed before it is completely burnt up in the fission process and it may then become associated with the fission products in their global distribution through the atmosphere and into rain. The cumulative deposition of Pu239 has been estimated to be about one-hundredth that of Cs137 with some values of about 2 mCi/km^2 in northern temperate zones.

Atomic weapons have also contributed to environmental contamination from Pu239 in a very different way involving accidents of strategic bombers carrying the weapons in readiness for a surprise attack. Two such accidents have been reported, the first occurring over southern Spain in January, 1966 when a US B52 bomber collided in the air with a refuelling tanker aircraft. Three bombs fell near to the village of Palomares and a fourth into the sea. The bomb is designed in such a way that two or more sub-critical masses of the fissile material are forced very rapidly together into a single critical mass by a conventional explosive before the nuclear explosion occurs. At Palomares there was no nuclear explosion, but the Pu239 was dispersed over the countryside by the conventional explosive in the case of two of the bombs which fell into the local tomato-growing fields. The bomb which fell into the sea was eventually recovered intact after a difficult and titanic struggle.

Another incident involved a bomber crash on the coast of Greenland leading to some contamination of sea-water but no evidence of serious widespread contamination of food.

A second isotope of plutonium, Pu238, produced in a nuclear reactor is used for a very different purpose as a power source for SNAP (Space Nuclear Auxiliary Power) devices in orbiting satellites and spacecraft. The energy from the radiations emitted by the Pu238 is transformed into heat which is then converted into electricity by the thermo-electric effect. Such devices known as nuclear batteries which can employ other radioactive sources, e.g. Sr90 should provide steady and reliable sources of electrical energy for many years. The re-entry of a navigational satellite into the stratosphere above the Indian Ocean in April 1964 led to its burn up and fairly widespread dispersal of the Pu238 in the southern hemisphere. 17kCi of the isotope were involved with maximum depositions in a belt between latitudes 20° and 40°S well below the fall-out of Pu239.

Plutonium, because of the nature of its decay scheme (it is an α-emitter similar to radium), is extremely hazardous if it is ingested and retained in the body. The movement of the radioelement through the biosphere and into food chains is however different from that of caesium and strontium described earlier. Compounds of the element and especially its oxides are characterised by their refractory and very insoluble nature. Their transfer into plants from the soil is therefore even less likely than caesium and strontium, and the extent of their transfer to dietary sources may be extremely small. The main problems with plutonium concern the refractory small particles which may be inhaled and lodged in the lungs or may be deposited and adhere on to plant surfaces. The incident at Palomares was, however, still sufficiently alarming to require the removal of all top soil from a considerable area of the fields around the scene of the crash.

RADIOACTIVE CONTAMINATION OF THE AQUATIC ENVIRONMENT

THE AVERAGE deposition per unit area of strontium-90 and caesium-137 over the oceans appears to be higher than the average deposition per unit area over land by a factor of between 1.5 and 3.0. The edible products of the seas however contribute only a small fraction of the total strontium-90 and caesium-137 in the diet, partly because of the enormous dilution factor of the oceans and partly because fish contribute only a small part of the total protein consumption by man except for a few small communities.

The average concentration in sea-water due to world wide fall-out may be aggravated in a few localised areas around the coast where regular discharges of radioactive waste are authorised. In the UK the highest permitted discharges occur off the Cumberland coast into the Irish Sea by pipeline from Windscale Works (Table 4.4). They are low level fission product wastes from parts of the nuclear fuel chemical processing plant. Monitoring of the sea-water, edible fish and seaweed in the area and elsewhere around the coast is undertaken regularly by the Fisheries Radiobiological Laboratories, Lowestoft.[2] Plaice are the principal species of fish caught commercially in the north Irish Sea; results from the region of the pipeline discharge suggest that the fish are able to concentrate the caesium-137 in their flesh by a factor of about twenty times the sea-water concentration. The concentration factor (CF) in this case can be expressed as follows:

$$CF = \frac{\text{activity due to Cs137 per kg. flesh}}{\text{activity due to Cs137 per litre of seawater}} = 20$$

Despite this concentration factor the average contamination level in the fish sold to the public is not regarded as at all serious, owing to the rapid dilution and dispersal of effluent from the end of the pipeline.

There is also some concentration of Cs137 in the edible seaweed Porphyra umbilicalis, which is harvested from the Cumberland coast and processed into laverbread for consumption by mining communities in South Wales. Another fission product however, ruthenium-106, Ru106 (see Table 4.2), is of far greater significance as a source of contamination for the miners of South Wales. The half-life of this radioisotope is fairly short so that the fission product is only of significance in the discharges to sea or from fresh fall-out. In either case but more especially in the case of the coastal discharges it is the special ability of many seaweeds to concentrate the Ru106 from sea-water and in the case of porphyra this concentration factor may be as high as 1000. This could readily lead to hazardous levels in the edible parts of porphyra if the discharges at Windscale were not in fact closely controlled and regulated by the amounts of Ru106 in the effluent. The levels in the final laverbread are further reduced by the incorporation of porphyra which is harvested from areas free from contamination.

The example of Ru106 in porphyra and laverbread affecting a small part of the population illustrates a critical pathway which is of special local significance. A variety of critical pathways affecting small population groups may also be found whenever active liquid effluents are discharged into a local environment, and their identification and study then becomes a matter of local importance. The discharge of cooling pond waters from the Bradwell Nuclear Power Station on the Blackwater Estuary in Essex provides a further example of a local contamination problem. The critical radioisotope in this case is an activation product Zn65 (Table 4.2), which may also be present in fall-out. Its presence in effluent arises from the corrosion of the magnox fuel cans during their immersion in the cooling pond. Zn65 in the fall-out which occurs over dry land and the resultant food contamination are quite negligible compared with marine pollution and especially a critical pathway involving the oyster. In the Blackwater estuary the oyster is the critical organism because there are oyster beds downstream from the discharge point and because of the high concentration factor of Zn65 in the edible flesh which may be between 1000 and 10,000 times as much as the sea-water concentration of the radioisotope. These concentrations appear to have fallen in recent years, possibly as a consequence of effluent treatment and the risk of human exposure from this critical pathway is very small. Further examples of critical pathways involve Cs137 and shrimps in the Severn Estuary at Hinckley Point and off-shore at Dungeness, but neither contributes very much to human contamination.

Edible foods from the freshwater environment will be even less significant as a source of human dietary contamination than marine produce. Although three establishments are permitted to discharge very low levels of radioactivity into rivers, the resultant contamination of drinking water supplies is quite insignificant. A rather special problem arises however at one Nuclear Power station in Britain. The station at Trawsfynydd is unique among the power stations in being the only one which is sited inland and on the shores of a freshwater lake in North Wales. It is the only station to use a freshwater coolant and to discharge low levels of radioactive waste into an inland lake. The critical pathway in this case was found to involve trout which are apparently much prized by the local anglers. The trout were found to concentrate Cs137 which is present in the discharge and in the fall-out from atomic weapons. The concentration factor in the flesh of the trout was found to be about two. Cs137 appears to have created a few local problems of this type through its ability to migrate to a small extent from the mixed fission products in the uranium fuel and to penetrate the magnox cladding. The problem at Trawsfynydd has been controlled by introducing a special effluent treatment process and reducing substantially the amount of Cs137 in the discharges into the lake. Local problems of this nature which are additional to the worldwide fall-out from weapons tests do serve to emphasise the necessity for thorough investigation and effective controls whenever limited discharges of radioactive effluent are permitted into the environment.

HUMAN CONSUMPTION OF RADIOACTIVE CONTAMINANTS

IN THE preceding sections strontium-90 and caesium-137 have been given special attention mainly in view of certain physical characteristics such as a long half-life which determines their persistence, their production by nuclear fission in high yield, and their solubility in water which is an important factor in their relative mobility and transfer through vital food chains. In this and later sections we are concerned with the final stages of their transfer to man, their distribution within the human body, and the significance of the radiation dose which they deliver to body tissues. Their hazardous nature to man is underlined by their close chemical similarity with the two elements calcium and potassium which are essential mineral elements in a healthy diet.

Animal produce such as milk and meat together with cereal grains and their products have already been mentioned as the principal human foods containing these two fission products. Although they are certainly responsible for a large proportion of the total intake of the two fission products by man, all foods are involved to some degree. The estimates of the average daily intakes of the two fission products are presented in Table 4.5. The data which have been used in compiling these estimates are based on the food analyses reported by the Letcombe Laboratory of the Agricultural Research Council,[1] and have been arranged in the seven principal food groups as used in the Total Diet Study for Pesticide Residues (*see* Chapter 1, p. 17). The food group contamination levels are given in the strontium and caesium units (SU and CU) as defined at the end of the chapter. These results and their relation to the body contamination levels are summarised in Table 4.6. The period 1963–4

Table 4.5
Sr90 and Cs137 in average UK mixed diet 1964[1]

(1) Food	(2) kg/day	(3) SU in food	(4) Consumption Ca % of total	(5) Consumption SU in diet	(6) CU in food	(7) Consumption K % of total	(8) Consumption CU in diet
Dairy produce including fats	0·45	27·4	59	16·1	63	32	19·4
Meat (Fish)	0·21	56·0	2	1·1	148	13	19·2
Cereals (70% extraction flour)	0·25	17·4	23	4·0	125	7	8·8
Root vegetables	0·26	48·4	2	1·0	14	16	2·2
Leafy vegetables and	0·10	30·6	5	1·5	6	32	1·9
Fruit and preserves	0·26	—					
Miscellaneous (drinking water)	—	—	9	2·1	—	—	—
Totals	—	—	100 (1·1 g)	25·8	—	100 (3·1 g)	51·5

NOTES

1. The total daily consumption of calcium is 1·1 g. The figures in column 4 represent the percentage of this total contributed by the various food groups.
2. The number of SU in each food group contributed to the total diet is the product of the number of SU in the food, column 3, and the percentage calcium, column 4, expressed as a decimal fraction.
3. The total daily consumption of potassium is 3·1 g. The figures in column 7 represent the percentages of this total contributed by the various food groups.
4. The number of CU in each food group contributed to the total diet is the product of the number of CU in the food group, column 6, and the percentage potassium, column 7, expressed as a decimal fraction.
5. About 28% of the total calcium in the average diet consists of mineral sources of the element – 22% as craeta praeparata or chalk in white bread and flour plus 6% as hardness in drinking water.
6. The fairly close agreement between the contamination levels of Sr90 expressed as SU in milk and in the total diet, and also those due to Cs137 expressed as CU in milk and in the total diet should be noted. This agreement means that monitoring of milk supplies can provide a useful guide to the contamination levels in the whole diet, and simplify the surveillance procedures.

Table 4.6

Summary of dietary intakes in 1964

Element or Radioisotope	Daily consumption	Body levels
Natural potassium	3·1 g	150 g
Natural K40	0·003 μCi	0·13 μCi
Cs 137	157 pCi	0·03 μCi
CU	51·5 pCi/gK	154·5 pCi/gK
Natural calcium	1·1 g	1000 g
Sr90	28·4 pCi	6500 pCi
SU	25·8 pCi/gCa	6·5 pCi/gCa

produced the highest contamination levels in the human diet; these levels have declined subsequently following fairly closely the reductions in annual fall-out over Britain and also the average contamination levels as shown in Figs. 4.7, 4.8.

METABOLISM OF THE RADIOACTIVE CONTAMINATION
(a) Strontium-90 and caesium-137

ONCE THE Sr90 and Cs137 have entered the body they will pass readily into solution in the stomach and move rapidly through the gastro-intestinal tract into the bloodstream. The Sr90 will find itself again in competition with its chemical analogue calcium and a certain fraction will follow the calcium into freshly forming bone. As in the case of the transport mechanisms through the food chains there will again be some discrimination against the Sr in favour of Ca as shown by the following observed ratios.

$$OR \text{ (bone/diet)} = \frac{SU \text{ in bone}}{SU \text{ in diet}} = \tfrac{1}{4} \text{ for adults}$$

$$= \tfrac{1}{3} \text{ for children}$$

The strontium concentrations in the other body tissues are much less than in the bone, which is therefore regarded as the critical organ for this particular fission product. The presence of Sr90 in milk and milk products is of special importance for young children, since in the case of those born since the 1950s the whole of the mineral bone will contain a proportion of this fission product. Measurements of Sr90 in infant bones have been regularly assessed by the Medical Research Council and a selection of these results are presented in Fig. 4.9.[7] Once again the levels are shown to vary in a similar fashion to the annual depositions and cereal concentrations. The Sr90 is held in the bones for a relatively long time as the turn-over of mineral elements in bones is slow.

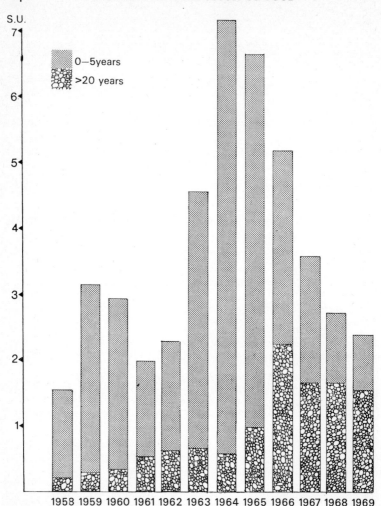

Fig. 4.9 Strontium-90 in human bone based on the MRC Monitoring Reports[7]

Cs137 which enters the bloodstream is in competition with potassium, and any Cs137 absorbed into the body tissues will tend to behave similarly to that element. Unlike Sr90 it will be fairly uniformly distributed in all tissues of the body and like potassium it will probably enter into all the cells which go to make up the variety of tissues. Whereas the radiation dose for Sr90 is delivered to the bone and bone marrow, that from Cs137 will be delivered to the whole body, including the critical gonad and bone marrow tissues. This is partly on account of its whole body distribution but also because it emits penetrating γ-radiations when it disintegrates, unlike Sr90 which is purely a β-emitter with the radiations having only a short range in bone and bone marrow. Cs137, like potassium, is also mobile in the body and there will be a fairly rapid turnover, but not quite so rapid as that of potassium. It is therefore found that there is an accumulation of Cs137 in the body relative to potassium as shown by the following Observed Radio (OR):

$$\text{OR (tissue/diet)} \; = \; \frac{\text{CU in tissue}}{\text{CU in diet}} \; = \; 3$$

It has to be emphasised that the contamination levels referred to in Table 4.6 are average national values for the whole of the population in Britain. They provide a reliable guide to the exposures of the majority of the population, but some variations from these values are to be expected for special groups, quite apart from individual variations. Levels in excess of these values will apply to the special groups of the population living in hilly districts having higher than average rainfall and fall-out; differences may also apply to other groups with dietary habits different from the normal dietary and to small groups living close to a major nuclear installation.

The results for foods from the special areas of the country are excluded from the national averages determined by the Letcombe Laboratory.[1] The average values for milk are based on a number of sampling stations covering a substantial part of the total milk production but with only relatively small differences in the contamination levels. The special areas which are subject to higher fall-out and have the poorer quality soils have however been covered by the surveys carried out by the Laboratory and as is to be expected the milk values in 1963–4 were greater than the national average; in the Cumberland area for example the values reported for milk were as follows:

Sr90, from 75 to 175 SU
Cs137, from 250 to 600 CU

As all the food which is produced in the special areas is subjected to the same higher fall-out the total contamination levels of the local population subsisting mainly on the local produce may be estimated to have been several times greater than the national average.

The additional contamination of the foods produced and consumed in the vicinity of a nuclear installation is more difficult to determine, but as it will affect mainly the produce of the seas the total consumed is not expected to be very much greater than the national average; it may also involve the special isotopes and critical pathways around the establishment.

Variations in dietary habits are not expected to affect the contamination levels very seriously except in the case of whole grain products substituting for flour of 70% extraction. Whole grain has a higher content of both Sr90 and Cs137 and the flour which is produced from it does not include the mineral calcium supplement which contributes almost one-quarter of the total calcium in the average diet. The Sr90 content expressed as SU is therefore several times greater than that of white flour and the value for the whole diet is about twice that of the average diet. In the case of Cs137 the higher contamination level is matched by a much higher potassium content in the whole grain so that the number of CU is less than that for white flour. These differences in the cereal products consumed are more significant than other dietary differences; the author has calculated that the average vegetarian or vegan diet is not significantly different in terms of the Sr90 contamination levels and is almost certainly less in terms of Cs137 contamination.

(b) Other radioisotopes

OF THE other radioisotopes referred to above (pp. 134–6), exposure of the whole population results from the increased fraction of carbon-14 in the carbon content of the food consumed and occasional exposures result from releases of iodine-131. Additional exposures from zinc-65, ruthenium-106 and a few other radioisotopes are restricted to special groups of the population in the vicinity of major nuclear installations except at times of fresh atomic fall-out.

Some increases in human exposure are also to be expected from the radioactive inert gases such as argon-41 and krypton-85 the levels of which will tend to increase with the expansion of the nuclear power programme. These gases are chemically inactive and will therefore remain in a chemically uncombined state and will disperse quite rapidly into the atmosphere adding very little to human radiation exposure in the immediate future.

The greater content of C14 in the human diet, about 80% greater in 1964 than the natural level, is incorporated along with the mass of stable carbon which is present in all the essential nutrient components of the diet including the structural and physiological proteins, and all the fats and carbohydrates which provide the source of the body's energy. The whole body content of C14 is therefore raised above the natural levels of this radioisotope, reaching a maximum increase of about 70% round about 1965. This increased level of C14 will also appear in the sugars and the organic bases which are vital components of the DNA in the cell chromosomes.

In view of the short half-life of iodine-131 exposures from this isotope are generally of short duration and confined to periods immediately following the testing of nuclear weapons or a nuclear incident, such as the Windscale accident in 1957. At other times, as at present, contamination levels in food and in the body are not detectable. Iodine-131 will be a key isotope in the early stages of any major nuclear accident and will determine the emergency actions required. As in the case of strontium and caesium, milk is a major dietary source of this isotope. Any contamination due to iodine-131 will be rapidly absorbed in the g.i. tract, the radioisotope associating with stable dietary iodine, which is an essential element in small amounts. This isotope is of special importance in view of its preferential concentration in the small thyroid gland of the body. About one-quarter of the total iodine absorbed from food is rapidly transferred to the thyroid gland, a fact of special significance for young children in view of the very small size of the gland. The highest general public exposures to iodine-131 occurred after the massive series of weapons tests in 1961–2 but exposures after the Windscale accident were restricted by withdrawing temporarily supplies of milk produced in the area around the plant.

Ruthenium-106 is similar to plutonium in its metabolism and is not readily absorbed through the alimentary tract, which is therefore the critical organ for any radiation exposure occurring during the passage of the food through the digestive system.

HUMAN EXPOSURES FROM RADIOACTIVE FALL-OUT

THE ACTIVITY levels and distributions of the various radioisotopes within the body will determine the magnitude of the internal radiation doses to the various tissues. These tissues vary in their sensitivity to damage by radiation and it is the doses to the more sensitive tissues or critical organs which are of prime concern. The critical organs for man are regarded as the bone marrow and the gonads (the testes and the ovary). Sr90 is incorporated in newly forming bone and the bone marrow is very sensitive to any radiations reaching it from the deposited radioisotope. 1 SU in the bone delivers an annual dose of 0·55 mrems to adult bone marrow; the average value of 6·5 SU in 1964 would have delivered a total dose of about 4 mrems in the year; the corresponding dose to the bone marrow of children would have been about 50% greater, at 6 mrems. The dose to the newly forming bone in either case would have been just under 20 mrems for the year. Cs137 by contrast with Sr90 is distributed throughout all the body tissues and both the gonads and the bone marrow are critical organs for the radiations emitted by this radioisotope. In this case 1 CU in the soft tissues of the body delivers an annual dose of 0·03 mrems; 155 CU in 1964 would, therefore, have delivered a dose to gonads and bone marrow of about

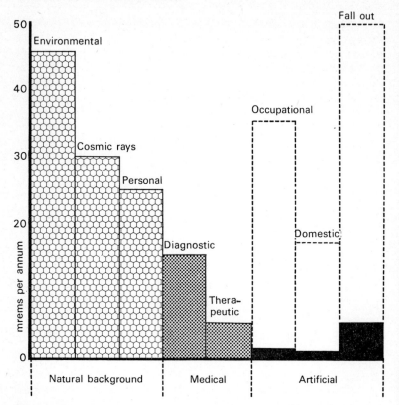

Fig. 4.10 Average human exposures to radiation in Britain in the late 1960s (see also Table 4.7 for a more detailed analysis of the background exposures).

NOTE

(a) The values for the occupational exposures are averaged out for the whole population.

(b) The dose limits for artificial exposures are represented in the figure by the heights of the dotted rectangles, and are based on values recommended by the International Commission on Radiological Protection. These dose limits assume that any increase in exposure entails some additional risk of deleterious effects which has to be balanced against any benefits which may be conferred. The maximum levels for occupational exposure of a small group ensure a negligible degree of risk of somatic injury, and the dose limits for the whole population minimise the risk of subsequent genetic damage.

4·5 mrems in the year. During 1964 at a time of comparatively fresh fall-out other fission products were also contributing to human internal exposure with the result that the average total exposure to bone marrow and gonads from all sources was about 15 mrems in the year. Subsequently the decline in the fall-out rates of Sr90 and Cs137 and the disappearance of the shorter lived fission products has led to a reduction in the annual exposure rate to about 3 mrems. There will also be a small addition to any internal dose from fall-out activity lying on the ground and contributing to an external body dose. The annual exposure rates due to fall-out are compared with all other sources of human exposure, natural and artificial in Fig. 4.10 and dose commitments estimated to the year 2000 are summarised in Table 4.8.

Table 4.7

Average natural background exposures to radiation

| | Source | Body tissue, doses in mrem p.a. | | |
		Gonads	Bone	Bone marrow
EXTERNAL	Cosmic radiation	30	30	30
	Environmental	50	50	50
	Total	80	80	80
INTERNAL	K40	20	15	15
	Rb87	0·3	< 0·3	< 0·3
	C14	0·7	1·6	1·6
	Ra226	–	6·0	0·3
	Po210	3·0	21·0	3·0
	Total	24	44	20
EXTERNAL + INTERNAL	Grand total	104	124	100

Table 4.8

Estimates of dose commitments to the year 2000 from radioactive fall-out and of the risks to the general public[8]

| Radioisotope | Estimated dose commitment (mrems) to the general public in the period 1954–2000 from all tests to 1965 | | |
	Bone	Bone marrow	Gonads and other tissues
Sr90	140	40	–
Cs137 (int + ext)	50	50	50
C14	31	13	13
Misc.	17	17	17
Total	238	120	80

See note on next page

The natural incidence of all forms of cancer is currently about 2,500 per million of the population per annum. Assuming that a radiation dose of 1 rem per annum will increase the normal rate of incidence by 2%,[9] this would result in an extra 50 cases per million of the population p.a. after a period of some years. The dose commitment reported in the table to the soft tissues of the body, about 100 mrem, over the period of almost 50 years is equivalent to an average annual dose of about 2 mrem. This might, therefore, be expected to increase the annual natural rate of incidence of all forms of cancer by 0·10 per million of the population after the delay period. The number of additional cancers in the 55 million population in the UK over a period of 50 years might be expected, therefore, to amount to about 275 compared with the natural incidence of about 6 million. The detection of such an increase will not be possible. The figure is probably not very meaningful serving only to emphasise the minute extent of any effect, although still implying the need for caution when all additional sources of exposure are considered.

HAZARDS TO MAN FROM IONISING RADIATIONS

THE ACUTE hazards to man and all other forms of life from massive doses of radiation are well enough known from the Japanese atomic bomb casualties and from a few accidental exposures which have occurred since 1939. It is therefore possible to suggest a provisional $LD_{50}/30$ for man of about 500 rems delivered in a short time to the whole body. (The $LD_{50}/30$ is the semi-lethal dose, i.e. the dose of radiation that will prove fatal after a period of 30 days to half of a large group of people receiving the dose.) The initial effects of such an exposure are generally felt within about two days as a general malaise with loss of appetite, nausea, vomiting, fatigue, and prostration. This is normally followed by a period of temporary recovery lasting up to 2–3 weeks, followed by a period of much more severe malaise, with fever, haemorrhages, and diarrhoea, lasting up to 6–8 weeks. This is the period during which the majority of the fatalities will occur, due mainly to severe damage to the bone marrow. Gradual recovery of the surviving half of the population begins after this time and may take place over several months. When the whole body exposure is about 1000 rems survival is unlikely with deaths ensuing much more rapidly within a few days mainly as a result of severe damage to the g.i. tract. When the dose is in excess of 10,000 rems death is certain and is extremely rapid within a few minutes or hours, with severe damage to the brain and central nervous system.

These acute effects for radiation exposure are not at all relevant to any effects which may result from exposure of the population to present levels of radioactive fall-out. Some knowledge of the long-term effects of exposure to sub-acute doses is based on studies of people such as uranium miners, luminisers and early radiologists, exposed as a result of their work, of patients subjected to radiation therapy, and the Japanese survivors of the atomic bomb explosions. The latter have been extensively studied by the American Atomic Bomb Casualty Commission which has been operating in Japan since the end of the Second World War.[10] The commission has found evidence for an increased incidence of various forms of cancer, which reached a maximum in the exposed population after a latent period, which is about 7–8 years for leukaemia and about 15 years for the other forms of cancer. Similar results are reported from the studies on radiologists and patients and all establish some sort of a relationship between the incidence of an effect (measured by the number of cases in a given number of the population) and the magnitude of the dose received (Fig. 4.11). All the known effects of ionising radiations occur quite

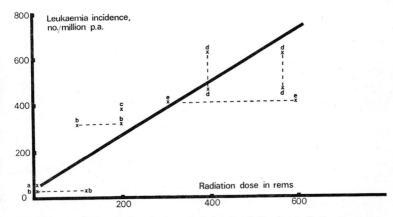

Fig. 4.11 Available data on leukaemia induction following radiation exposure, after J. C. Collins (personal communication).[11] The letters represent data from the following sources:

a.   The natural leukaemia incidence rate, i.e. 50/million population p.a.
bb. The Japanese atomic bomb survivors
c.   X-rayed children.
dd. Adult patients treated for ankylosing spondylitis (stiff back)
ee. American radiographers

In the case of bb and ee, the data on leukaemia incidence are firmly established but there are considerable uncertainties in the dose received. In the case of dd the leukaemia data is uncertain but the dose received is known precisely. The line which has been drawn represents a fit to the data assuming that there is no leukaemia threshold dose

naturally and spontaneously in the human population with radiation possibly just another factor hastening the onset of the symptoms. There are no specific effects which are unique to radiation unless a change in the sex ratio i.e. the ratio of male to female births in the normal population, or a reduction in the normal life-span as a result of non-specific ageing processes can be included.

The effects of ionising radiations have probably been the subject of more intensive studies than any other form of environment pollutant. The chronic effects of human exposure and especially the genetic consequences of exposure have been backed up by massive series of experiments involving small experimental animals such as mice and insects such as the fruit fly. The latter has been found invaluable for assessing the genetic effects of exposure, which in view of the basic similarity of cells in all organisms must have some relevance for the human situation. Extrapolation of all such data to the conditions of human chronic exposure is however fraught with considerable difficulty but it has been established beyond question that all ionising radiations are (1) carcinogenic when body tissues such as the bone marrow, the bone, thyroid, and lung are exposed; and (2) that they are also mutagenic when the gonad tissues are the recipient of the dose and teratogenic when the foetus is exposed especially in the early sensitive period when organogenesis is at its most rapid. Most of these facts are based on incontravertible evidence, but again it has to be emphasised that all the investigations have again involved levels of exposure which are still well above those which are prevailing from worldwide radioactive fall-out. Any genuine assessment of the effects at these extremely low levels of exposure may well have to be related to knowledge of the precise mechanisms involved in the chain of events from exposure to effect.

Alternatively some guidance may be obtained eventually from studies of small population groups who are exposed quite naturally to significantly higher levels of background radiation. These studies embrace the citizens of La Paz, Bolivia living at an altitude of about 12,000 ft., and receiving a much higher cosmic radiation dose, and the people of Kerala in India living on beds of radioactive monazite sands and receiving a much higher environmental dose. No evidence is available at the moment to suggest any difference attributable to the extra doses.

As in the case of a great many agents responsible for human contamination, the chronic long-term effects of exposure to very small doses of radiation can only be imperfectly deduced from knowledge of effects at very much higher doses. The ultimate effects of exposure to radiation may not be very different from the effects of many chemical agents; the absorption of radiation energy is a physical interaction with body tissues, imitating a sequence of chemical changes. One of the primary effects of ionising radiation in tissue therefore is to produce a number of reactive chemical agents which may change the chemical nature of biological molecules or seriously interfere with the complex cycle of molecular processes within the cells of the tissue. Reactive chemical species produced by the physical interaction of the radiations are the essential precursors for any changes within the cells and ultimately within the tissues. This sequence of events – physical, chemical, biological, is probably enhanced in tissues owing to the high water content of all body tissues. Most of the radiation energy is absorbed by the water, giving rise to several highly reactive chemical products of water, which in turn react with the molecular contents of the cells. A small number of chemical changes may result from the direct interaction of the radiations with cell molecules but this process is unimportant compared with the indirect process involving reactive species in the water.

By contrast with many chemical agents of pollution however the radiations emitted by radioactive substances have considerable powers of penetration so that the actual damage may be remote from the site occupied by the substance and any part of a cell is likely to be a target for radiation damage. Chemical agents on the other hand have to surmount many barriers in the course of their transport through the body and their effects are usually limited to specific receptor sites. Although the precise sequence of events leading to cell damage from exposure to radiation is still imperfectly understood, what is known is sufficient to suggest that certain changes may well be irreversible and may occur at any level of exposure, including background exposure.

It is undeniable that radiation is an important causative factor in cancer induction and genetic mutations; mutations occurring in the body cells may indeed be a prior stage to cancer induction. It is unlikely, however, that the quantitative data on the incidence of any form of cancer due to radiation alone can be established with certainty until the precise sequence of biological events leading to the induction of the cancer are known. A further and possibly more serious source of error in arriving at estimates of radiation damage is associated with the multiplicity of agents known to be involved in cancer induction. The ability of many chemicals, including a number of drugs in common medical use, to damage bone marrow cells is well known; the bone marrow may then be sensitised to damage by radiation which is then a further factor in the chain of events leading to leukaemia. It may, therefore, be misleading to attribute the incidence of all cancers even in cases of known exposure solely to the radiation factor when other primary factors may also be involved. In these cases estimates of damage due solely to radiation may be too high.

Despite all the uncertainties and in view of the need to pursue a cautious policy, especially where the future health of new generations is concerned, it is obviously wise to adopt the pessimistic conclusions arrived at by extrapolating the data from high dose rates. This is the policy adopted by the International Commission on Radiological Protection in making their recommendations for maximum levels of radiation exposure to the general public.[12] The estimates for the incidence of damaging effects from the present levels of fall-out have been arrived at in this way and are summarised in the note to Table 4.8.

Whereas it is possible to arrive at estimates of the possible damage to human health resulting from atomic weapons fall-out, the problems are considerably more complex when the hazards of a developing nuclear power programme have to be evaluated. The magnitude of any risk to the human population depends to some extent on the type of nuclear power reactor to be installed at the station, the siting of the station, and the nature of the fuel processing cycles which will be required. The chances of a major accident at a nuclear station leading to a large-scale release of fission products, are extremely small in view of all the safeguards built into the reactors. There is none the less a small finite risk and the probability of an accident occurring must therefore increase with the number of stations which are commissioned. In the event of such an accident leading to a release of activity immediate problems will be created by the radioisotopes of iodine. The effects of these radioisotopes can be offset to some degree by the consumption of tablets containing stable iodine as iodide or iodate which will effectively dilute the radioactivity due to the iodine fission products and block to some extent the entry of these products into the thyroid gland. These tablets are available for issue at all the major nuclear stations in the UK, and additional protection can be afforded by restricting the consumption of locally produced foodstuffs, and more especially milk. As a last resort, temporary evacuation of the population from the immediate area may be called for. Later problems may well be created by other fission products, and especially the more persistent $Sr90$ and $Cs137$. The problems will be essentially similar to those which have been outlined for the fall-out from atomic weapons, although the levels of activity locally may greatly exceed the highest contamination levels for these two radioisotopes experienced so far from atomic weapons.

In normal operation there will be some increase in radiation exposure of the general public from the continuing discharges of the radioactive inert gases, and some increase in local exposures from the discharges of liquid wastes into the local environment. The latter may involve a critical pathway leading to some food consumed by a small number of the local population, as shown on pp. 127, 137. The problems of general public exposure from the expanding nuclear energy industry in the USA have recently received a great deal of attention in that country as a result of the forceful representations of Drs Gofman and Tamplin.[9] This has led to a considerable reduction in the radiation levels permitted for the general public in the vicinity of these stations. The problems of the expanding nuclear power industry in the UK have also been reviewed in this country by the recently established National Radiological Protection Board.[13] Despite the difficulties in accurately evaluating the magnitude of the risk to the public, this is certain to be extremely small, and may be the inevitable price which has to be paid for ensuring supplies of electrical energy in the future. It is clearly imperative that the small groups of the population in the immediate vicinity of a nuclear station should not be exposed to any substantially greater risk compared with the rest of the population simply as a result of the circumstances bringing a nuclear power station to their doorstep.

The major problems of an expanding nuclear power industry are much more likely to concern the long-term storage of the large quantities of highly radioactive wastes, and also the security of the growing stocks of the artificial fissile element plutonium. These security measures are becoming increasingly necessary to eliminate the risk of nuclear high-jacking, through, for example, a threat to a communities water supplies or of a nuclear explosion by terrorists.

NOTES ON RADIATION UNITS

(a) **The Curie, Ci,** is the unit of source strength, and is a measure of the rate at which a source is disintegrating and hence of the rate at which the radiations are being emitted.

1 Ci is an activity of $3 \cdot 7 \times 10^{10}$ disintegrations/sec (dps)
i.e. 37,000 million dps.

1 Ci is, therefore, equivalent to an activity of
$2 \cdot 22 \times 10^{12}$ disintegrations per minute (dpm)
i.e. 2·22 million million dpm.

Atomic weapons release megaCurie (MCi) quantities of fission products in a test explosion, 1 MCi being a million Curies. Dispersal of this activity in the air and into rainfall leads to considerable dilution, and the following sub-multiple units then become relevant.

1 mCi = 10^{-3} Ci (one-thousandth of a Curie) = $2 \cdot 22 \times 10^9$ dpm.
1 μCi = 10^{-6} Ci (one-millionth of a Curie) = $2 \cdot 22 \times 10^6$ dpm.
1 pCi = 10^{-12} Ci (one-million-millionth of a Curie) = 2·22 dpm.

The mass of a radioactive isotope associated with a given activity is often extremely small. 30 pCi of strontium-90, for example, has a mass of only 0·2 pg; it has $1 \cdot 33 \times 10^9$ atoms in this weight, of which about sixty disintegrate every minute, each disintegration producing a β-ray.

(b) **The caesium unit** (CU) is used to express the activity of caesium-137 in foods and other materials relative to their content of potassium.
1 CU = 1 pCi caesium-137 per g of potassium.
The strontium unit (SU) is derived in a similar way for strontium contamination levels but relative to calcium content.
1 SU = 1 pCi Sr90/g of calcium.

(c) **The rem** is a unit of biological dosage. It is a measure of the amount of radiation energy absorbed in the tissues but weighted with quality factors to allow for a non-uniform distribution of the energy absorption in the tissues.

The average annual background dose of radiation to the critical organs, the gonads and bone marrow, is 100 mrem (Table 4.7).

One SU delivers a dose of 3·7 mrem per annum to bone and 0·55 mrem per annum to adult bone marrow. The dose to the bone marrow of children is higher at 0·82 mrem per annum.

One CU delivers a dose of 0·03 mrem per annum to body tissues.

(d) **Half-life**. The half-life of any radioisotope is the time, expressed in years, days, hours, minutes, or even seconds, during which the initial activity of a source is reduced to one-half. The half-life is an almost unique characteristic of a radioisotope but there is a very wide range of values.

e.g.

Uranium-238	=	4·5 billion years, i.e. $4·5 \times 10^9$ years
Carbon-14	=	5760 years
Radium-226	=	1620 years
Strontium-90	=	28 years
Ruthenium-106	=	1 year
Iodine-131	=	8 days

REFERENCES

[1] Annual Reports of the Agricultural Research Council Letcombe Laboratory (formerly known as the Radiobiological Laboratory), which have appeared at regular intervals since 1959; published by the Agricultural Research Council, HMSO, London, 1959 to 1972.

[2] Technical Reports of the Fisheries Radiobiological Laboratory, Lowestoft, FRL 2, 5, 7 and 8, N. T. Mitchell, Ministry of Agriculture, Fisheries and Food, HMSO, London, 1968 to 1972.

[3] *Radioactive Fall-out in Air and Rain: Results to the Middle of 1971*, R. S. Cambray *et al.*, Health Physics and Medical Division, AERE, Harwell, Berks. AERE–R6923. HMSO, London, 1971.

[4] *Environmental Radioactivity in Denmark in 1968*, A. Aarkrog and J. Lippert, Danish Atomic Energy Commission Research Establishment Riso, Riso Report No. 201, Roskilde, Denmark, 1969.

[5] "Caesium-137 in Plant Produce, II Distribution of Cs137 in the Soils and Produce of an Experimental Farm", R. Perry and J. W. Lucas, *Plant Foods for Human Nutrition*, 1972, **2**, 193.

[6] *The Hazards to Man of Nuclear and Allied Radiations*, a 2nd Report to the Medical Research Council, Cmnd 1225, HMSO, London, 1960.

7 *Assay of Strontium-90 in Human Bone in the UK*, Medical Research Council Monitoring Report Series, a series of reports from 1960, HMSO, London.

8 Report of the United Nations Scientific Committee on the effects of Atomic Radiation, General Assembly Official Records: 17th Session Supplement No. 16 (A/5216), United Nations, New York, 1962 (and several later reports).

9 *Population Control through Nuclear Pollution*, A. R. Tamplin and J. W. Gofman, Nelson-Hall, Chicago, 1970.

10 Reports of the Atomic Bomb Casualty Commission, Hiroshima, Japan.

11 J. C. Collins, unpublished figure, University of Manchester Radiological Protection Service, Manchester.

12 Recommendations of the International Commission on Radiological Production ICRP 9 adopted 17th September, 1965, published for ICRP by Pergamon Press, Oxford, 1966.

13 *Living with Radiation: a booklet for the layman*, National Radiological Protection Board, HMSO, London, 1973.

Suggestions for further reading

The Assessment of the Possible Radiation Risks to the Population from Environmental Contamination, a Report to the Medical Research Council by their Committee on Protection against Ionising Radiations, HMSO, London, 1966.

Radioactive Fallout, Soils, Plants, Foods, Man, ed. Eric B. Fowler, Elsevier Publishing Co., Amsterdam, New York and London, 1965.

Radioactivity and Human Diet, ed. R. Scott Russell, Pergamon Press, Oxford, 1966.

5

Industrial Pollution by the Mineral Elements

INTRODUCTION

THE EARTH'S crust and oceans contain about ninety naturally occurring elements, of which about seventy are basically metallic elements and the other twenty essentially non-metallic elements. The latter include the very abundant and vital elements of all living matter, carbon, hydrogen, oxygen and nitrogen, with smaller amounts of other elements, such as chlorine, sulphur and phosphorus. A few other non-metallic elements at very low concentrations, other halogens and selenium, are also known to have essential roles within living organisms, but they may also become hazardous at higher concentrations. Generally, the non-metallic elements do not present serious health hazards to man through their ingestion in food, although a few problems may be encountered from elements such as arsenic and these are reviewed briefly at the end of this chapter.

Very many more of the metallic elements are, however, likely to be involved at potentially hazardous concentrations in food, largely due to their increasing exploitation by man. Very many of these elements are known to be essential for healthy growth in plants, animals and man at low concentrations and only become hazardous when certain threshold concentrations are exceeded. The essential elements for man have been defined by Spivey-Fox as elements "specifically and irreplaceably required for some direct metabolic effect that is mandatory for survival or beneficial to the health of the organism".[1] The essential elements include those which are present at macro-concentrations both in the diet and in the body, e.g. sodium, potassium, calcium, and magnesium, and a great many others which are only present in the diet and in the body at much lower concentrations. These latter range from the moderately abundant elements in the diet with well-established roles, such as iron, zinc, copper, and manganese, to other elements such as tin, nickel, vanadium, which appear to be beneficial to health but whose precise role has not been identified (Table 5.1). The number of elements listed in the table is only a small fraction of the total number of terrestrial

Table 5.1

Average normal concentrations of some essential metallic mineral elements in the environment, in the human diet and in man, after H. J. M. Bowen[2a]

Element	Average Natural occurrences Soils (ppm dry wt.)	Oceans ppm	Approx. normal food intake mg/day	Approx. normal human levels[2b] Whole body ppm	Blood mg/100 g
Macro-elements					
Sodium (Na)	6300	10,500	2000	1500	200
Potassium (K)	14000	380	3000	2000	170
Calcium (Ca)	13700	400	1100	15000	6
Magnesium (Mg)	5000	1350	400	500	4
Trace elements					
Iron (Fe)	38000	0·01	15	57	50
Zinc (Zn)	50	0·005	15	33	0·9
Copper (Cu)	20	0·003	5	1·4	0·1
Manganese (Mn)	850	0·002	5	0·3	0·003
Chromium (Cr)	100	0·00005	0·1*	<0·1	0·003*
Cobalt (Co)	8	0·00027	0·0002	<0·05	0·00003
Molybendum (Mo)	2	0·01	0·7*	<0·07	0·0004
Tin (Sn)	10	0·003	40	0·43	0·01*
Nickel (Ni)	40	0·0054	0·5	0·14	0·004
Vanadium (V)	100	0·002	0·3*	<0·002	0·002*

*Provisional figures

metallic elements, but excludes all those for which a need has not been experimentally demonstrated. Many of these latter elements do occur in the normal diet, in some cases at concentrations greater than those of the essential elements but so far without an identifiable role. As Bowen has suggested, however, it could be quite wrong to regard these elements as non-essential and only a very few elements, mainly radio-elements, which in some cases are present in tissue cells at concentrations of one atom per cell or less, may justifiably be so regarded. The toxic effects of a number of non-essential metallic elements, more especially cadmium, mercury and lead, have claimed attention in recent years as have a number of the essential elements such as copper, nickel but at higher concentrations than normal. The case of lead has attracted considerable publicity and is dealt with separately in Chapter 6. Mercury has also been responsible for a number of outbreaks of food poisoning and will be discussed in some detail in the present chapter with reference to its chemical analogues, zinc and cadmium.

There is one important distinction between the mineral elements as a group and some of the contaminants discussed in the previous chapters, more especially the organic chemicals used as pesticides and radioisotopes produced in the processes of nuclear fission or activation. As stable chemical elements they have a permanence which is not exhibited by these other contaminants; organic chemicals undergo slow or rapid biotransformations which change the great majority of them into totally innocuous substances or in just a few cases they may be changed for a time into more toxic metabolites. Artificial radioisotopes all undergo radioactive decay which again may be rapid or slow depending on the half-life of the active material, but resulting in most cases in the formation of a stable and totally harmless product. The mineral elements are more akin to the primordial radioelements described in Chapter 4 having been incorporated in the earth's crust at the time of its formation about 4·5 billion years ago. No process will change the amounts of these elements in the total environment although changes in their chemical state may alter the degree of their toxicity and their exploitation by man may alter their normal distribution within the environment. These elements are processed into many forms for the range of their applications and the chemical processing wastes are dispersed into the biosphere. The net result is a continuing process of transfer of the mineral element from its normal state in the lithosphere into a modified state in the biosphere; these processes are also taking place naturally as a result of the forces of erosion; human activities are merely speeding up the natural processes of dispersal and are accelerating the problems of contamination.

There is always a degree of natural contamination due to the mineral elements as a consequence of these natural processes; it would have been invaluable to have had data on their natural distributions in foods and in man prior to their large scale exploitation more especially since the period of the Second World War. The problem is relatively simple in the case of artificial radioisotopes and organochlorine pesticides as there is no natural occurrence of these substances and any changes taking place progressively in the biosphere can readily be assessed and related to human activities. The position is very different for the mineral elements however as there is generally insufficient background data to distinguish any effects due to human activities from the perfectly natural effects.

ZINC, CADMIUM, AND MERCURY

THE THREE elements zinc, cadmium, and mercury form a fairly closely related group of elements within the periodic classification of all known elements (Appendix). They are also sometimes grouped with another much more closely related family of elements, the alkaline-earth elements calcium, strontium, and barium. Although all six elements do have some properties in common, the differences between the more important alkaline-earths and the zinc group however outweigh any resemblances they may have. There is a marked increase in human toxicity with increasing atomic weight from zinc to mercury; the chemical properties of mercury are often fairly unique and distinctive from those of zinc and cadmium. This is illustrated especially by the fact that mercury forms a series of extremely stable organic compounds which can be very persistent in the natural environment and are a great deal more toxic than the inorganic forms of mercury. The organic compounds of zinc and cadmium are very much less stable and the toxicity of cadmium is almost certainly due to inorganic forms of the element. All three elements, however, share the ability to associate closely with proteins, mainly through their ability to link with the sulphydryl (-SH) groups of amino-acids in the protein chains, and this has important physiological consequences.

Cadmium is the 67th element in order of abundance and normally occurs in fairly close association with zinc, with a cadmium to zinc ratio of 1:350 by weight in most of the mineral sources (Table 5.2).

Table 5.2

The natural occurrence of zinc, cadmium, mercury

Element	Earth's crust	Concentration ppm		Coal
		Soils	Oceans	
Zinc	70	50	0·005	40
Cadmium	0·2	0·06	0·00011	0·25
Mercury	0·08	0·03	0·00005	–

In view of the naturally close association between zinc and cadmium occasional cases of zinc poisoning may, in fact, have been due to the associated cadmium rather than to the zinc. It is convenient therefore to consider these two elements together and to treat mercury as a separate problem.

ZINC AND CADMIUM IN THE ENVIRONMENT

IT CAN be seen from the table that zinc is a relatively abundant element (25th in order of abundance); there are a few concentrated ores, such as zinc blende (sphalerite, ZnS) which is often found in association with dolomite and calcite. By contrast, there are very few mineral sources of cadmium, the most important being Greenockite (CdS); most of the cadmium which is extracted and used by man is obtained as a byproduct from zinc refining. Some cadmium is also found in association with lead and it has been reported that superphosphate fertilisers may also sometimes contain quite high concentrations of cadmium. World production of zinc is running at about one million tons per annum and of cadmium about 15,000 tons per annum. Zinc finds a great variety of applications, being particularly well-known for its use in galvanised steel and in dry batteries. The element is also a component of many alloys, including brass, its sulphide is used as a pigment in paints, and the salts find many uses. Cadmium is used as a component of various low melting point alloys, such as silver solder, in bearing alloys, in electro-plating for protective coatings and also in batteries. The sulphide is also used as a pigment in paints and plastics.

The total quantities of the two elements reaching the environment as waste-products and their critical pathways through the environment which culminate in food and human exposure are not known with any certainty. As in the case of mercury, however, it is expected that transport of the two elements through marine food chains will assume major importance for food contamination especially in coastal waters. This follows from the high accumulation factors of the two elements in marine plants, fish and shellfish, and also from the fact that the concentration levels of the two elements in coastal waters appear to be related fairly sensitively to the quantities discharged in industrial wastes. Some analyses of coastal waters in the Irish Sea to the west of England have been carried out and the results of the surveys are presented in Table 5.3.

Table 5.3

Zinc and cadmium in coastal waters[3]

Area	Zinc ppm			Cadmium ppm		
	Mean	Highest	Lowest	Mean	Highest	Lowest
Liverpool Bay	0·012	0·048	0·002	0·00027	0·00074	0·00014
Cardigan Bay	0·0075	0·020	0·004	0·00111	0·00241	0·00048
Bristol Channel	0·010	0·021	0·004	0·00113	0·00420	0·00028

The rather higher concentrations in Liverpool Bay and the Bristol Channel compared with the average values of Table 5.2 are mainly due to industrial discharges but the higher concentrations in Cardigan Bay

Table 5.4

Concentration factors of zinc and cadmium in marine plants and fish at normal sea-water concentrations, based on Bowen[2a]

Element	Average sea water conc. ppm	Concentration factor	
		Marine plants	Fish
Zn	0·005	30,000	1200
Cd	0·00011	3,650	1350

are largely due to natural erosion and run-off from the mineralised region of Snowdonia.

The concentration factors of the two elements in marine produce may be inferred from the average concentrations based on the data presented by Bowen and are shown in Table 5.4. The maximum concentration of cadmium in the Bristol Channel area, about forty times the normal average concentration in sea-water, could evidently lead to some quite high contamination levels in any fish landed from the area.

By contrast with the high concentration factors in the marine environment, the corresponding values from soil to plants and animals are quite small, mainly as a consequence of zinc and cadmium being firmly bound to a component in the soil, probably clay or organic matter. The concentration factor from soil into plants for zinc is estimated to be about two and that for cadmium about ten.

ZINC AND CADMIUM IN FOOD

AS IN the case of many elements, foods provide the largest sources of zinc and cadmium intake for human beings. Both elements have been detected in urban atmospheres, typical concentrations being about 50 μg per cubic metre (50 μg/m^3) of zinc and 0·1 μg/m^3 of cadmium. Despite these concentrations it has been estimated that the intake by the inhalation of urban air is less than 1 % of the intake from food.

There are no regular monitoring programmes for the analysis of zinc and cadmium in foods which are at all comparable with those which have been organised for pesticide residues and radioactive contamination. One of the most comprehensive surveys of concentrations in food was carried out in the USA by Henry Schroeder and his colleagues and a selection of their results is given in Table 5.5. Using this data as a

Table 5.5

Examples of zinc and cadmium concentrations in selected foods[4]

Food Group	Item	Concentration ppm Zn	Cd
Cereals	Whole wheat	31·5	0·25
	Flour, white	8·9	0·27
	Cornflakes	76·5	0·39
Meat and fish	Beef	56·6	0·89
	Lamb	53·3	3·49
	Pork	3·6	0·32
	Fish*	17·5	0·79

Table continued on next page

Food Group	Item	Concentration ppm Zn	Cd
Fats	Butter	2·0	0·56
	Lard	–	0·05
	Margarine	1·55	0·8
Fruits		0·5	0·04
Root vegetables	Potatoes	8·7	0·03
	Turnips	12·1	0·03
	Carrots	5·2	0·30
Green vegetables	Greens	0·41	0·06
	Spinach	2·21	0·45
	Cabbage	1·82	0·07
Milk and dairy produce	Wholemilk, canned	4·2	0·10†
	Milk, dried, non-fat	35·1	0·49

* The average concentrations in fish exclude any contribution from oysters which may contain zinc up to 1500 ppm and cadmium up to about 4 ppm. The cadmium concentration of fish is reduced below the quoted average value if anchovies, 5·4 ppm, are also excluded.
† Provisional figure

guide and the average food consumption levels in the UK, the total dietary intake of the two elements is estimated to be about 15 mg zinc and 0·35 mg cadmium (Table 5.6). These estimates compare very well

Table 5.6

An estimate of the daily dietary intake of Zn and Cd

Food group	Consumption % w/w	Average concentration ppm Zn	Cd	Consumption mg/day Zn	Cd
Cereals	16	19·5	0·18	4·9	0·05
Meat, fish	13·5	29·1	0·8	6·1	0·17
Fats	3	1·5	0·50	0·1	0·03
Fruits	13	0·5	0·04	0·1	0·01
Root veg.	17	3·5	0·07	0·9	0·02
Green veg.	7	1·5	0·2 *	0·1	0·02
Milk	29	4·2	0·1 *	1·9	0·05
Misc.	1·5	–	–	–	–
Total	100% (1·55 kg/day)	9·0	0·25	14·6	0·35

*Provisional figures

with further results of Schroeder based on the analysis of a USA institutional diet and with the values of 10 to 15 mg zinc and 0·6 mg cadmium quoted by Bowen.[2a] Estimates based on Schroeder's data, however, appear to be somewhat higher than those reported in some other USA surveys, and the average normal daily intake may, therefore, be assumed to range from 10 to 15 mg zinc and 0·1 to 0·6 mg cadmium. It will be noted that animal produce and cereal grains contain the highest concentrations and provide the major dietary sources of these two elements for the average population. Higher intakes may be expected in the case of small groups of the population consuming larger quantities of fish from coastal waters or consuming fresh produce from farms close to industrial sites, and possibly from produce grown from soils fertilised with superphosphate. These increases, which are expected to be small, are unlikely to be significant in terms of human health, especially in the case of zinc.

ZINC AND CADMIUM IN MAN

ZINC IS an essential element with a daily requirement which must be close to its normal intake of about 15 mg per day but no human need for cadmium has, as yet, been demonstrated. Both elements are absorbed fairly rapidly from the g.i. tract (about 50% of the intake in the case of zinc) and are then transported by the bloodstream to the liver, kidneys and other body tissues. Experiments with animals suggest that most of the unabsorbed zinc and cadmium is eliminated in the faeces, with a smaller quantity in the urine and small losses in sweat. There appears to be a slow accumulation of the two elements in the body up to about the fifth decade of life, with a preferential accumulation in the kidney tissues and especially in the renal cortex, although the blood concentrations appear to attain a fairly rapid equilibrium with the amounts

Table 5.7

Concentrations of zinc and cadmium in body tissues[2b]

Tissue	Concentration ppm Zn	Cd
Whole body	33	0·43
Blood	9·3	<0·01
Bone	66	N.D.
Brain	13	<0·9
Kidney	48	32
Liver	46	2·44
Muscle	50	<0·7
Testes	15	<0·6
Prostate	87	—
Hair, nails	250	—
Eyes (choroid, retina)	<500	—

N.D. = Not detectable

ingested daily. The accumulation in the kidney is associated with the presence of a special protein, thionein, in the cortex which is able to bind the two elements strongly. The normal distribution of the two elements in various body tissues is shown in Table 5.7. The whole body contains about 2 g of zinc, which is a similar quantity to that of iron, and almost two-thirds of this amount is found in muscle tissues and about one-fifth in bone. High concentrations are found in hair, nails, the choroid of the eye and the prostate. The total body burden of cadmium is about 30 mg, about one-third being present in the kidney and one-sixth in the liver with only small amounts in bone.

The ratio of the amounts of cadmium to zinc in soils (about 0·2 μg cadmium per mg zinc) increases in the average human diet to about 26 μg cadmium per mg zinc, and then decreases in the whole body to about 13 μg cadmium per mg zinc. The ratio in the kidney tissues is, however, extremely high at a value of 667 μg cadmium per mg of zinc. It will be noted that white flour has a much higher cadmium/zinc ratio (about 40 μg Cd/mg Zn) than has the whole grain (about 8 μg Cd/mg Zn) which should ensure a rather smaller cadmium intake when wholemeal flour and bread are preferred to white flour and bread. A similar enrichment has been reported for polished rice in comparison with the unpolished whole rice.

In view of the essential human need for zinc, it is most improbable that foods could ever contain sufficient of this element to constitute a toxic hazard. Some reports of cases of "zinc poisoning" arising from the use of galvanised utensils for acidic foods have probably been due to the cadmium content of the zinc. Zinc has an essential role to play in promoting and controlling several vital metabolic processes in the body, a role which is becoming increasingly apparent. Its activity in the body is due to its association with certain protein enzymes. In some cases it is intimately bound with the protein to form a number of metallo-enzymes, of which about ten have been identified to date. It is also an activator of several other enzyme systems. Zinc is also regarded as beneficial in wound healing and in controlling hardening of the arteries. High concentrations of zinc have also been found in liver tissues adjacent to tumours and are believed to represent a defence reaction to the invasion by malignant cells. In all probability, then, zinc deficiencies are more likely to present health problems than excessive amounts. These deficiencies have, in fact, been associated with dwarfism and enlarged livers in some Middle East populations. Exceptionally large intakes of zinc in food at about 5000 ppm, compared with the present human dietary levels at about 9 ppm, have been found to cause some adverse effects in experimental mice, including growth inhibition, anaemia (possibly due to iron displacement from the heme proteins) and a reduction in reproductive capacity. For man himself the 15 mg of zinc consumed per day at a concentration of 9 ppm in the food, must be regarded as a normal requirement and it is not possible to define at all accurately an upper limit beyond which toxic effects might become evident.

It is, however, highly probable that cadmium concentrations in excess of the normal food concentrations of 0·25 ppm are far more significant for human health. At higher concentrations cadmium is almost certainly in competition with zinc and may inactivate those enzymes which incorporate or are activated by zinc for their proper functioning. It is to be regarded, therefore, as an anti-metabolite of zinc. The adverse effects on small experimental animals, such as anaemia and stunting of growth and impairment of reproduction, occur at very much lower diet concentrations than for zinc. There is also some evidence that cadmium is responsible for kidney damage and that it might be a factor in increased blood pressures, with the increase roughly proportional to the ratio of cadmium to zinc in the food. Bowen quotes a figure of 3 mg per day representing a concentration of about 2 ppm in the diet as toxic to man. There appears to be no conclusive evidence, however, linking cadmium as a carcinogen, mutagen, or teratogen for man, although further experimental studies have been suggested.

One authentic outbreak of cadmium poisoning, referred to as the Ouchi-Ouchi disease, has been reported, and affected a small community near Toyama City in western Japan. A total of about 56 deaths are reported to have occurred over a number of years as a result of the consumption of rice and soya beans contaminated by cadmium to a concentration of about 3·5 ppm. from local mining and industrial operations. The daily intake of cadmium was estimated to have been about 1·3 mg. The cause of death was variable but the clinical symptoms included rheumatic pains, osteoporosis, increased urinary excretion of calcium and amino-acids, and disturbed kidney function.[5]

ZINC AND CADMIUM EFFECTS ON COPPER

THE ELEMENTS iron, cobalt, nickel, copper, and zinc are close neighbours in the periodic chemical classification of the elements, and are known as transitional elements. Cadmium although it is in a higher period of the classification table also has fairly close affinities with this first series of elements, all of which appear to have a vital functional role in a healthy organism. Copper and zinc are especially close neighbours, and they are similar in their chemical behaviour, with the result that there are physiological interactions between the two elements, which lead to competition between them for absorption from diet and also in their ability to associate with certain proteins as enzyme co-factors. Levels of zinc in food, which are in excess of normal dietary requirements may depress the absorption of copper resulting in copper deficiencies, and vice versa. Copper has been shown to be an essential trace element with anaemia as one well established effect of copper deficiencies in the diet. The element is very widely distributed in the environment and typical concentrations are reported by Bowen, and presented in Table 5.8. The normal average food consumption is 2 to 3 mg per day, with a total body content of about 100 mg, a body concentration of between 1·5 and 2·5 ppm. About one-seventh of the total amount

Table 5.8

Average concentrations of copper in the environment, ppm[2a]

Marine environment		Terrestrial environment	
Sea-water	0·003	Soils	20
Plants	11	Plants	14
Molluscs Crustacea	50	Animals	2·4
Fish	4		

is found in the liver at a concentration of 7 ppm, and there is a concentration of about 100 $\mu g/100$ ml in blood.

Copper salts are sometimes used as growth promoters in intensive stock rearing, especially of pigs, so that somewhat higher than normal levels of copper may be expected in some animal produce, e.g. > 50 ppm in lamb liver. There is no evidence that there is any serious increase in copper consumption from food resulting from any human activity which is sufficient to create any problems of zinc metabolism and utilisation or which might lead to any symptoms of copper poisoning. More serious problems may well arise from copper deficiencies either from reduced levels in the diet or from competitive absorption resulting from higher than average levels of zinc. It has also been amply demonstrated by experiments involving farm animals that higher than average concentrations of cadmium in the diet can interfere with copper metabolism leading to reduced levels of copper in the liver and in the blood.[6] These deficiencies of copper have been associated with effects such as impairment in the growth of the animals and in their reproductive performance. All the evidence would appear to suggest therefore that cadmium is a more powerful competitor with copper than is zinc, not only for absorption from diet but also in disturbing those enzyme reactions in which copper is an important co-factor.

The experiments on animals are not relevant to any human situation resulting from increases of cadmium or zinc in the average diet, as they are carried out at concentrations well above any human dietary concentrations likely to be encountered. Nevertheless, this ability of cadmium to compete with copper and to interfere significantly may well be the most important consequence of higher concentrations of cadmium in the diet. There is, however, one possible compensating factor in that all three elements tend to appear at higher concentrations together in the environment as in the case of pastures close to sources of industrial contamination. The extent of any enhanced concentrations of the three elements in the diet will then depend particularly on the chemical forms of the contamination and the availability of the elements to the grazing animals.

MERCURY AND ITS COMPOUNDS

MERCURY, or quicksilver, has been known to man and used by him since antiquity. Its long history is partly due to the ease with which the element can be extracted from its principal ore, cinnabar, but it is none the less a rare element with an overall abundance in the earth's crust of about 80 ppb establishing it sixteen places from the bottom of the list of all elements arranged in order of their natural abundance. It is a unique element being the only one which is a liquid metal at normal temperatures and its unusual properties, such as its ability to dissolve other metals forming various amalgams, have always ensured its share of attention by man. It was for many years a key element in many early theories of matter and in processes used by the alchemists; its use in medicine also has a long history. There are still several entries in the *British Pharmacopoeia* which include the element and some of its compounds. One of the compounds with chlorine, mercuric chloride (corrosive sublimate), is a very effective antiseptic, while the finally divided metal in the form of "Blue Pills", and another compound with chlorine, mercurous chloride (Calomel), are used as purgatives. There has also been a long awareness of the toxicity of some of these inorganic compounds of mercury but the additional threat to man and his environment of certain organic forms of mercury has only become apparent in quite recent times, although it has always existed naturally to a small extent. The growth in the production and applications of mercury, and especially of a number of organomercurials used as fungicides, has aggravated this natural situation.

As is the case for most elements, mercury is very widely distributed

Table 5.9

Average concentrations of mercury in the environment, ppm[2a]

Marine environment		Terrestrial environment	
Sea-water	0·00003	Soil	0·1
Plants	0·03	Plants	0·015
Fish	0·12	Animals	0·046
		Coal	0·03

in the natural environment but at extremely low average concentrations (Table 5.9). There are very few mineral sources of the element, the principal one being the compound with sulphur, cinnabar, small deposits of which occur in Spain, Italy, parts of North America and the Far East. The metal is very readily extracted from the ore by distillation in air and can also be readily converted into the variety of compounds which find many applications. The estimated annual world production of the element is about 10,000 tons of which approximately 50% finds its way back into the environment, mainly into surface waters and into the oceans. This rate of annual transfer to the biosphere from industrial processes and agricultural applications, is matched by a similar amount which enters the oceans of the world by the continual erosion of natural sources.

The production of mercury in the United Kingdom is about one-tenth of the world production and the main applications are shown in Table 5.10. The low abundance of mercury, the small number of available

Table 5.10

Estimated usage of mercury in the UK[8]

Application		Estimated percentage Annual usage
Chemical industry	1. Mercury cells in the chlor-alkali process	49·0
	2. Pharmaceuticals	1·8
	3. Plastics manufacture	1·8
	4. Paint formulations as a fungicide	7·0
Electrical industry	Switchgear, batteries	35·0
Agriculture	Fungicides – seed dressings	3·5
Miscellaneous	Laboratory instruments, detonators, polishes and waxes, wood pulp and paper	1·8

first class ores, with the extremely large and rapid increase in its utilisation in industry in the post-war years are matters of anxious concern as the known global resources of the element are now being used up at such a high rate that an available lifetime of only 15 years has been forecast.[7]

Mercury cells in which the mercury is used as a liquid electrode in the electrolysis of brine are widely used in industry to meet the growing demands for the production of chlorine and also of caustic soda. In most countries these represent the largest use of mercury and they are also a major source of fairly widespread artificial contamination of the environment. Although measures to conserve mercury in such plants and to reduce losses to the environment are constantly being improved, some losses are unavoidable. Some of the mercury is released into the atmosphere with the hydrogen gas also produced on these plants and other losses occur by drainage to surface waters. A comprehensive study of mercury pollution in the Swedish environment[9] showed that a substantial fraction of the element reaching the biosphere from human activities came from the chlor-alkali industry, with much smaller contributions from the wood-pulp and paper industries using phenyl mercury acetate (PMA) as a slimicide, and all the other miscellaneous uses including the use of mercurial seed dressings in agriculture.

The harmful effects of mercury resulting from all these activities which add to the natural sources of mercury entering into the biosphere only first became apparent in the latter part of the 1950s, when the element was identified as the cause of severe epidemics of food poisoning in the Minamata Bay area of Japan. These outbreaks of severe mercury poisoning resulted in 43 deaths out of 111 cases of reported poisoning over a period of about ten years, and deaths are still occurring. The poisoning was eventually traced to excessive concentrations of mercury, up to 50 ppm in fish and shellfish caught in the bay and consumed by the local population. The source of the high concentrations was found to be the effluents from industry and the local paper mills, which contained both mercury and large quantities of organic matter; reaching the sea-water together, these encouraged the growth of bacterial organisms with an ability to convert the mercury regardless of its original chemical form into the very persistent dimethyl mercury, which is soluble in the water and is concentrated by the fish. The human population was not the only form of life to suffer from this chain of events as the local cats and sea birds were also quite seriously affected.

The Japanese investigations into the Minamata poisonings prompted other investigations into the presence and effects of mercury within the environment. A large-scale investigation into the declining populations of game birds was started in Sweden, and also covered other effects of mercury poisoning. The practice of dressing grain seeds with organo–mercurials, prior to sowing, was found to be an even more important source of damage to the bird population than the organochlorine pesticide residues described in Chapter 2. Concentrations in excess of 2 ppm were frequently found in the livers of seed-eating birds, their predators, and also in the eggs of pheasants. Comprehensive surveys of the distribution of mercury in the Swedish environment were carried out in 1967, as a result of which the government banned the consumption of fish from 40 lakes, where mercury concentrations in the fish were found to exceed 1 ppm. The ban on the consumption of some fish in Sweden was followed in the early 1970s by further bans imposed on the consumption of canned tuna and swordfish in the USA. There was also a move to replace the more toxic forms of mercury used as seed dressings by less toxic forms of mercury or by other types of fungicide altogether. The further dangers of applying mercury compounds to the dressing of cereal grain seeds were also emphasised by severe and acute outbreaks of mercury food poisoning reported from Iraq, Guatemala, and in Pakistan in the 1960s, when cereal grains so treated were accidentally diverted into milling and flour production for human consumption. A further outbreak in Iraq in 1971, resulting in many casualties and deaths has also been reported.

MERCURY IN THE ENVIRONMENT

THE 5000 tonnes of mercury in industrial wastes matched by the equivalent amount from natural processes of erosion, enter the environment mainly via surface waters creating special problems in coastal areas. A very small fraction of the total quantity is released to the atmosphere; in the case of human activities this arises mainly from the chlor-alkali plants, about 20 tonnes per annum in Sweden, with smaller amounts from the combustion of coal and paper wastes. Small quantities are also reported to be released naturally in volcanic eruptions and this has been confirmed recently in the eruption on Heimaey Island off the coast of Iceland. Atmospheric mercury will lead to some fall-out and deposition on to soils but only in areas close to chlor-alkali plants is this likely to be at all significant and perhaps approaching the levels which are found quite naturally in the soils. The contamination levels in soils are however increased by the sowing of grain seeds and seed potatoes dressed with mercury fungicides. The processes of mercury transfer from soils to plants and its entry into foodstuffs have not been investigated systematically but the availability of mercury in soils is likely to be small due to its tendency to bind with organic matter.

Mercury residues in potato crops produced from seed dressed with organo-mercurials have been measured; in one special investigation potatoes dressed initially with 100 ppm methoxy methyl-mercury and having concentrations at planting from 0·9 to 2·3 ppm gave a maximum concentration in the harvested crop of 0·03 ppm[10] only slightly greater than the average value quoted in Table 5.12 below.

The problems concerning the soil as a source of dietary mercury are however almost certainly minimal compared with the much more serious problems which originate from the quantities reaching freshwater lakes or the oceans and especially the seas close to the shore. The processes of natural erosion contributing about 5000 tonnes of mercury to the oceans every year will normally be well dispersed and diluted by the large volume of water, and the more serious problems will undoubtedly result from the localised discharges of industrial wastes. The aquatic food chains are therefore of major importance for mercury entering into food especially in certain localities, of which Minamata Bay provides a good but unfortunate example, and these food chains are therefore emphasised in Fig. 5.1.

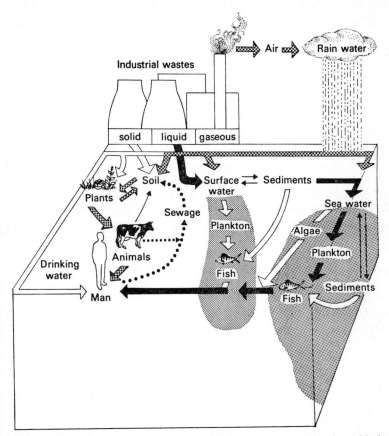

Fig. 5.1 Mercury in food chains. The black arrows represent the critical pathway for food contamination in man

Much of the mercury, whether as the element, or as a compound, which enters freshwater lakes, or sea-water close to the shore, appears to be fairly rapidly absorbed on to the bottom sediments and silts. There is some evidence that the mercury is at first held preferentially by particles of clay minerals in the sediments and silts. It is subsequently released into the water at a slow rate either before or after undergoing biotransformations by bacteriological action into methyl mercury compounds. These and other similar compounds are distinguished by their great stability, their solubility in sea-water, and their persistence in the aquatic environment. Mercury dimethyl and other similar organic compounds of mercury have been known to the chemist for over a century, and are quite readily synthesised in the laboratory. Their highly toxic nature has also been known for a long time, but knowledge of their natural origins in the environment and their entry into food chains is much more recent.

The total concentration of mercury in normal surface waters is extremely small, even when additional contamination from artificial sources occurs, but may still become sufficiently high to affect the aquatic environment adversely in two distinctive ways. Firstly, it has been demonstrated that many compounds of mercury can interfere seriously with photosynthesis and affect the growth of seaweeds and phytoplankton (see, for example, R. Nuzzi, *Nature*, 1972, **237** 38). It has been shown that the compound phenyl-mercuric acetate (PMA) when present at the concentrations of total mercury normally found in sea-water, inhibits the growth of several species of phytoplankton, which normally grow in sea-water, or in freshwater environments. The phytoplankton are a starting point of most food chains in sea-water assuming the role of photosynthesis normally carried out by plants in the soil. The possibility therefore exists that very small concentrations of organic mercury compounds in sea-water may have an effect on the productivity of the oceans, especially as PMA is rather less toxic than the methyl-mercury compounds. Similar effects have also been demonstrated in freshwaters, at organo-mercurial concentrations which are only ten times greater than the maximum concentrations of mercury in drinking water, as recommended by the USA Public Health Service. The second factor of concern in aquatic environments is the movement of these organic compounds of mercury into food chains and their increasing concentrations at each successive stages of the chain, from the simple phytoplankton to the fish consumed by man (Fig. 5.1). Probably the best known and a widely reported example of this concentration effect in a food chain concerns the predatory tuna fish which is a hunting fish feeding on other fish and crustaceans of the oceans. At the end of 1970 many batches of canned tuna fish in the USA were found to have high mercury concentrations, ranging from 0·3 to 1·12 ppm, fresh weight. Similarly high values were also found in some samples of canned swordfish.

A large proportion of the total mercury content of fish is present in the more toxic organic forms and not necessarily due to artificial contamination. The concentrations which have been found in samples of tuna fish and swordfish are rather extreme cases of the more normal contaminations found in fish, and they are quite small compared with the maximum values of 50 ppm which were found in the samples of fish and shellfish consumed by the local population in the Minamata Bay area of Japan. The tuna values are also quite high compared with the average levels of mercury contamination found in the fish landed in the United Kingdom, as reported by the UK Working Party on the Monitoring of Food Stuffs for Mercury and Other Heavy Metals.[11a] The concentrations of mercury in the fish landed in the UK vary with the distance of the fishing grounds from the ports of landing and range from an average of 0·06 ppm in fish from the distant fishing grounds to 0·11 ppm for the middle distances, and increasing to 0·21 ppm at distances of up to 25 miles from the coastline. Two-thirds of the total fish landed in the UK are from the distant fishing grounds and an average value for all the fish available to the consumer was found to be 0·08 ppm. There was little evidence that the contamination levels varied between the different species of fish, and the really important factor appears to be the distance of the fishing grounds from the ports. A few localities close to the shores of the British Isles were found to have even higher values of contamination as, for example, in the Thames Estuary with a maximum value of 0·45 ppm and in the Morecambe Bay area with a maximum value of 1·1 ppm. The fish from these areas and canned tuna fish imported into the UK represent only a very small fraction of the total fish consumed. Some concentrations in imported canned tuna were found to exceed 0·5 ppm in samples tested by the Government Chemist.[11b] Small groups of the population might therefore be exposed to above average intakes of mercury contamination from these particular sources.

MERCURY IN FOOD

FOOD UNDOUBTEDLY provides the most important sources of mercury for man with very much smaller contributions from the slighter contaminations of the atmosphere and drinking water. By comparison with the regular and systematic surveys of radioactive contamination in foodstuffs, the analyses for mercury are of much more recent origin. Two surveys of mercury in the average human dietary in England and Wales were based on the total diet studies of 1966/7 and 1970/1 which were organised by the Working Party primarily to determine the contamination levels of pesticide residues. The samples for analysis were portions of the composite dietary samples collected for these studies and were arranged in the seven major food groups after their preparation for consumption. These composite samples were collected from six large towns and cities in the United Kingdom.[11]

A summary of the results for the first quarter of 1971 is presented in Table 5.11, column 3. The results were similar to those obtained in the preceding and succeeding quarters with the one exception that the

Table 5.11

*Estimate of daily consumption of mercury in first quarter 1971,
UK[11]*

Food group	% w/w of total food consumed	Estimated average concentration of mercury $\mu g/kg$ (ppb)	Approximate daily consumption, $\mu g/day$
Cereals	16	5	1·3
Meat, fish	13·5	20	4·2
Fats	3	< 5	< 0·3
Fruits and preserves	13	5	1·0
Root vegetables	17	5	1·3
Green vegetables	7	10	1·0
Milk	29	< 5	< 2·3
Misc.	1·5	—	—
Totals	100·0 (1·55 kg)	< 8	< 11·5

average concentration in the meat and fish group was then only one-half
of the 20 $\mu g/kg$ shown in Table 5.11. The results shown in the table
for the first quarter of 1971 therefore represent the maximum values
for a quarter and justify the conclusions of the Working Party that the
average daily intake of mercury from food in the normal dietary is not
greater than 10 μg when averaged out over a year. These latest results
are not very significantly different from the previous estimate of an
average consumption of about 14 μg per day, based on the earlier 1966/7
dietary study. The results, taken altogether, suggest that there has been
no serious deterioration in the mercury contamination levels during the
latter part of the 1960s. Much earlier analysis of mercury in a variety of
foodstuffs were carried out in Germany and the USA and although
comparisons are difficult because of refinements in the analytical procedures and of national differences in food habits, they appear to confirm that there has been no very marked deterioration in the average
human consumption of mercury in recent times. Selected results for
individual foodstuffs, also organised by the Working Party, are compared with the results of the earlier analyses from Germany and the
USA in Table 5.12.

The results for the individual foodstuffs confirm the importance of
fish as a potentially serious source of mercury in the human dietary.
The consumption of fish in the average diet is, however, only a small
fraction of the total food consumption amounting to no more than
about 1·5%, which is about 25 g per day out of the total of 1·60 kg
consumption. The consumption of mercury from this source in the
average diet is estimated at about 2 μg on the assumption that the

Table 5.12

Mercury concentrations in individual foods

Foodstuff Group	Food sample	1970–1 APA*	1934–8 Germany†	1964 USA†
			Mercury concentration µg/kg (ppb)	
Dairy produce	Milk	10	0·6 to 4	8
	Cheese	170	9 to 10	80
	Butter	10	70 to 280	140
	Eggs	N.D.	2	
Meats	Beef	10		
	Pork	20		
	Liver, beef	40	1 to 67	1 to 150
	Canned meat	N.D.		
	Sausages, beef	40		
Fish	Salmon, canned	80		
	Shellfish	70	20 to 60	0 to 60
	White fish, fresh	70		
Cereals	Bread, white	20		
	Flour, white	20	20 to 36	2 to 25
	Breakfast cereals	—		
Vegetables	Potatoes	20		
	Salads	10	2 to 44	0 to 20
	Green vegetables	25		
	Peas, canned	20		
Fruit	Apples	10		
	Pears	N.D.		
	Tomatoes	10	4	4 to 30
	Tomatoes, canned	20		
	Dried	40		

N.D. = Not detectable

* Results of a survey of mercury in food by the Association of Public Analysts.[12] In most cases the results are average values for a small number of samples and provide a guide to contamination levels only. See also Report of the Government Chemist, 1971.[11b]

† L. J. Goldwater[13]

average level of mercury in the fish consumed is 80 µg/kg. Higher intakes of mercury up to the maximum of 50 µg per day could, however, apply to small groups of persons consuming larger quantities of fish, up to 100 g per day, at concentration levels up to 500 µg/kg and obtained from local fishing grounds close to the British Isles.

MERCURY IN MAN

THE BRITISH Working Party set up to investigate mercury levels in food and the total diet also arranged a small pilot survey involving a small number of analyses of mercury in the whole blood and hair of individuals representing special groups of the population. The results of these analyses are reproduced from the report of the Working Party in Table 5.13. A group of six people (C) were selected to represent the

Table 5.13

Mercury concentrations in whole blood and in hair[11]

		Sex	Age (years)	Mercury concentration (mg/kg) Whole blood	Hair
(A)		Consumers of fish from British coastal waters			
	1	M	34	0·01	2·1
	2	M	33	0·015	1·2
	3	M	62	0·03	3·9
	4	M	26	0·01	1·7
	5	M	50	0·01	2·1
	6	M	26	0·02	3·4
(B)		Consumer of canned tuna fish			
	7	M	17	0·015	2·4
(C)		Average dietary group			
	8	F	22	0·005	2·7
	9	F	27	0·005	2·2
	10	F	20	0·005	8·4
	11	M	58	0·005	0·8
	12	M	40	<0·005	1·0
	13	M	52	0·005	2·2

average diet in the UK and therefore served as controls for comparison with another group (A) who consumed a larger quantity of fish harvested from the coastal waters close to the British Isles. As it is to be expected, the second group have a greater daily intake of mercury from their food due to the extra fish consumption and this is also reflected in the higher concentrations found in the samples of whole blood and hair. The single individual, group B, consumed consistently larger quantities of tuna fish.

Although the sample survey of the different types of consumer was very limited it is possible to make some tentative estimates of the concentration factors from the diet to blood for the different groups:

Group A: The Fish Consuming Group

Concentration of mercury in diet $= 29 \, \mu g/1 \cdot 6 \, kg = 18 \, \mu g/kg$
(assuming 100 g fish having a mercury concentration
of 210 $\mu g/kg$ replaces 100 g meat)

Average concentration in blood $= 16 \, \mu g/kg$

Hence concentration factor $= \dfrac{\text{conc. in blood}}{\text{conc. in diet}}$ $= 0 \cdot 89$

Group B: The single tuna fish consumer
    Concentration of mercury in diet $= 18\,\mu g/kg$
    (assuming 100 g tuna fish having a mercury concen-
    tration of about 200 $\mu g/kg$ replaces 100 g meat)
    Concentration in blood $= 15\,\mu g/kg$
    hence concentration factor $= 0 \cdot 83$
Group C: The average dietary group
    Concentration of mercury in diet $= 10\,\mu g/1 \cdot 6$ kg $= 6\,\mu g/kg$
    Concentration of mercury in blood $= 5\,\mu g/kg$
    hence concentration factor $= 0 \cdot 84$

Although it is important not to place too much reliance on these estimates for the reasons stated and also for the assumptions made in
arriving at the dietary concentrations they do appear to show a consis-
tency with virtually no discrimination against the mercury in its passage
from the food to the blood. This might also suggest that the chemical
state of mercury in all the foods consumed in the total diet is similar
and that as the fish contain their mercury mainly in the methyl mercury
form this might also be the chemical form in the other foods. It should
be emphasised however that the results quoted in Tables 5.11 and 5.12
make no distinction between the organic and inorganic forms and
represent the total mercury content of the food.

As in the case of the other elements dealt with in this chapter the
lack of comparable data for earlier periods makes it impossible to
establish a datum line which could be used as a background level
representing the normal intake due almost entirely to the natural distribution of the element and its normal presence in food. In particular it
cannot be established that human intakes since 1945, with the rapid
expansion in the various applications of mercury, have shown any
significant increase for the general population. The results of the earlier
survey in Germany do lend some support to show that they have not
changed very drastically if at all, and that only in the case of the special
groups of the population subsisting on a much higher consumption of
fish can it be asserted that their mercury consumptions have almost
certainly increased.

Observations made mainly with experimental animals indicate that
the mercury which enters the body in the food consumed is mainly
absorbed through the g.i. tract, the extent and the rate of the absorption
depending on the chemical form of the mercury. Methyl-mercury compounds appear to be absorbed quite rapidly to the extent of about 90%,
whereas the inorganic forms are absorbed more slowly and only to the
extent of about 50%.

After absorption the organic forms of mercury appear to bind
preferentially to the red blood cells to be distributed fairly widely

throughout all the tissues of the body. Higher concentrations are found subsequently in certain tissues such as those of the liver, kidney, and brain, and these may be several times greater than the plasma concentrations. The organic compounds are eliminated relatively slowly mainly in the faeces, and probably involving the entero-hepatic cycle. The inorganic forms of mercury by contrast appear to be bound preferentially to the protein of the blood plasma and are then concentrated mainly in the kidney, which becomes the organ of elimination; the rate of removal appears to be faster than the organic forms. Examples of typical concentrations in various tissues of the human body are presented in Table 5.14, but they represent total values without regard to the chemical form.

Table 5.14

A guide to some typical concentrations of mercury in human tissues, compiled from various sources (See, for example, Holden A. V., Present Levels of Hg in Man and his enviroment, Chapter 12 Hg Contamination in Man and his Environment, IAEA Technical Report Series 137, Vienna 1972)

Tissue	Concentration, ppb
Blood	30
Muscle	20
Liver	100*
Brain	50*
Hair	1000 to 8000
Fingernails	650
Kidney	800
Urine	20

* Provisional figures

NOTE

The normal concentrations are associated with an average dietary intake of about 10 μg per day. Toxic symptoms probably begin to occur at a dietary intake of about 300 μg/day of methyl-mercury compounds, resulting in blood concentrations of about 200 ppb, a urine concentration of 100 ppb, and a steady state concentration in hair of 60,000 ppb.

BIOLOGICAL EFFECTS OF MERCURY

MERCURY IS almost unique among the various food pollutants as several serious incidents have occurred from the consumption of heavily contaminated food. Epidemics of acute food poisoning, referred to in the introductory sections of the chapter, have involved serious physiological effects and, in some cases, deaths. The outbreaks have been due either to the consumption of cereal grains dressed for sowing, but accidentally diverted to the production of food, or in the consumption of heavily polluted fish close to industrial discharge sites off the coast of Japan. The symptoms of acute poisoning have been described as severe abdominal pains with nausea, vomiting and diarrhoea, accompanied in some cases by severe damage to the kidneys and liver and disturbance of the nervous system. These latter effects on the central nervous system involve headaches, numbing of the tongue, lips and extremities, with irritability, bad temper and impaired coordination. It has been estimated that the concentrations of mercury, mainly in the form of methyl mercury compounds, in the fish and shellfish at Minamata which caused the 43 deaths from 1953 to 1966 reached values of 50 ppm. These high levels were also associated with above average consumptions of fish with a daily consumption of mercury amounting to about 5 mg over a period of time, which may be compared with the UK national average of about 10 μg per day and a maximum of about 30 μg from the total food consumed. Some of the organic compounds and, more especially the methyl derivatives of mercury, are particularly toxic to man and produce their acute toxic effects at lower intakes than the elementary or inorganic forms.

The difference in toxicity between organic and inorganic forms of mercury has also been confirmed by experimental work on small laboratory animals, exposed to chronic levels of the various compounds. The simple organo-mercurials such as the methyl-mercury compounds were found to be a great deal more toxic than the inorganic compounds, with certain organo-compounds such as PMA having an intermediate toxicity. Some of the chronic effects of organo-mercurials were found to occur at dietary concentrations of about 0·5 ppm, about one-hundredth of the concentrations found in some of the people at Minamata Bay.

The chronic effects of exposure to mercury are associated with its known ability to combine with the sulphydryl (SH) grouping found in a few of the amino-acids which go to make up body proteins. This property has been widely used for some time as mercury salts are a well-known analytical reagent for the separation of proteins from solution. Although there are differences in the strength of the binding of the different forms of mercury to protein, the process is responsible for the initial distribution to all body tissues either by attachment to plasma proteins or red cells. The mercury in combination is then enabled to penetrate to various parts of tissue cells, where it may interfere with the proper functioning of membranes or of functions mediated by the cell enzymes. This ability of mercury to bind to protein may also interfere with those liver enzymes which are normally involved in the detoxication of other foreign substances or in the potentiation of chemical drugs.

The nature and extent of these interactions has received very little attention to date.

The mechanisms responsible for the damage to the brain and other body organ tissues may also involve the same interaction of mercury with the sulphydryl group of enzyme proteins. The damage in this case is only apparent when a certain number of tissue cells have been affected; the damage due to organo-mercurial compounds owing to their great stability and retention in the tissue cells is likely to be irreversible and, therefore, cumulative. The damage may not become apparent until a certain threshold concentration has been built up in the tissue cells over a period of time. The permanent and more severe long-term chronic effects involving the brain and the central nervous system may cause mental and emotional disturbances accompanied by loss of concentration and of memory, general weakness, and fine tremors in the limb extremities and may lead eventually to paralysis, loss of vision and disturbed cerebral function; the toxicity thresholds for man which result in damage to the nervous tissues cannot yet be defined with any precision in the light of present knowledge.

There is also evidence that compounds of mercury may be a causative factor leading to long-term carcinogenic, mutagenic and teratogenic effects. The existing evidence for mutagenic effects has been reviewed by the Mrak Commission whose work has also been referred to in Chapter 2 on pesticide residues. Organo-mercurial compounds have been shown to produce chromosome aberrations in both plants and the fruit fly, drosophila. The evidence for man and other mammals, however, is not available although a limited investigation in Sweden appears to have demonstrated an association between chromosome aberrations and mercury levels in blood cells in the case of a small number of Swedish fish eaters. The Commission concludes that a considerable extension of the long-term mutagenic studies, especially for mammals, is imperative. A teratogenic effect has also been demonstrated in the case of experimental mice at an intake level of approximately 0·1 mg/kg body weight, resulting in foetal malformations, involving the eyes, tail and central nervous system. Certain teratogenic effects such as infantile cerebral palsy have also been observed in the case of a few children born to parents affected by the Minamata poisonings. Here again, there is a clear imperative for urgent further studies.

Despite a considerable literature on the toxic effects of mercury and several of its compounds, the United Kingdom Working Party is compelled to admit that there are "no long-term studies in animals and man which would attempt at present an estimate of the ADI for man"[11a] This is due to the fact that it is quite impossible at the present moment to establish any safe threshold value for the effects of poisoning or for the longer-term carcinogenic, mutagenic and teratogenic effects. With

rapid developments in science of the previous century and the growing confidence and ability of man to harness the results of basic science to modern technology. This acceleration in the rate of technological change, based on scientific discoveries, is illustrated well by the rapidly decreasing interval of time elapsing between an original discovery and its technological application. Light sensitive chemicals were discovered in the 17th century, but more than one hundred years were to elapse before photography became an established business. The principle of converting sound into electrical impulses was discovered in the 19th century and only about fifty years were required for its application to the development of the telephone service. In the 20th century, only an interval of five years was required to perfect the atomic bomb after the discovery of nuclear fission in 1939, and only a further ten to fifteen years were needed for the development of large scale nuclear reactors for generating electric power.

The second spurt in the Industrial Revolution followed on the end of the Second World War and witnessed the extremely rapid developments in nuclear technology, in the chemical industries, and in space technology. The emergence of the chemical industries was, indeed, a feature of the first Industrial Revolution. For the first time in recorded history, man then acquired gradually the ability to manipulate his chemical environment on an expanding scale. These large-scale developments of the chemical industries enabled the production of many natural chemicals for specifically human purposes, and these had previously only been present in the environment and interacted with man on a very modest scale. The production of artificial chemical fertilisers illustrates the rise to prominence of a small number of naturally occurring chemicals. The chemist is also able to create and to develop methods for producing a great variety of purely artificial chemicals having no natural occurrence and posing very different types of environmental problems. The manufacture of polymeric substances for the extensive range of modern plastics, many of the processes being based on an extension of the petroleum industries and the production of pesticide chemicals, such as DDT, are all excellent illustrations of this new capability. As a consequence a great many chemicals, either as intermediates or as end products of the chemical industry, were able to enter the environment and to create many of the modern pollution problems.

The great chemical expansion of the post-war years was also accompanied by even greater ingenuity in the production of an increasing range of synthetic chemicals, and an expansion in the fine chemical and pharmaceutical industries. A considerable number of these synthetics have been used as food additives at one time or another. It has, for example, been estimated that the average American is consuming about

5 lb of food additives every year, which is equivalent to a total consumption of about 500,000 tons per annum by the whole public. Several hundred of these substances are currently licensed for use in America and many of these have been introduced in the last 25 years. The actual number in use, at any one time, in significant quantities, will of course be considerably fewer. They are added to foods for a variety of reasons including preserving and extending the shelf-life, increasing their attractiveness by improving the appearance to the eye and taste, and maintaining a satisfactory texture. The chemicals which are added include the following groups of agents:

> Flavouring agents and condiments
> Colouring agents
> Preservatives, such as anti-microbials and oxidants
> Emulsifiers
> Sweeteners
> Vitamin and mineral supplements
> Flour improvers and modified starches.

Just as in the case of some of the environmental pollutants, some of the chemicals which are used are present quite naturally in foods, but others are entirely artificial and not normally present; many of them add nothing to the nutritional value of the foods.

The modern trend towards convenience and packaged foods is another factor which may also lead to contamination of food. Modern plastics are based on a number of long chain polymers, such as polyethylene and polyvinyl chloride (PVC) which are basically chemically inert and would not themselves contribute significantly to the contamination of food. The contaminating agents in this case are residues of chemicals employed in the polymerisation process or added subsequently to improve the appearance and the qualities of the plastic for the variety of uses to which the finished product is applied. These additives include chemicals employed as plasticizers, colouring agents, stabilisers to heat and light, fire retardants and anti-oxidants. It is these agents together with process residues, e.g. vinyl chloride monomer in the polymer which may migrate slowly into the foods from the packaging materials adding to the other chemicals which may already be present.

In addition to all these substances and the wastes of industry, other problems of human contamination are created by the pharmaceutical industries with their increasing ability to modify natural drugs to increase their specificity in medicine or to create entirely new synthetic drugs. Some indication of the magnitude of this problem is also illustrated by data from the United States. In 1966, for example, it is reported that 167M prescriptions were issued for a range of medications which included stimulants, sedatives, analgesics and tranquillisers. In a small survey, conducted in 86 households in San Francisco, there was an average of 30 medicaments per household of which about one-fifth were obtained on prescriptions. It has also been estimated that about 9M doses of amphetamines and about 40 tons of barbiturates were also being produced annually in 1962. The problem with these particular chemical drugs and others, such as LSD, is that probably about half of these quantities are entering into the illicit channels of distribution. (See, for example, *Drugs of Abuse*. ed. S. S. Epstein, M.I.T. Press, Cambridge, Mass., 1971.)

(b) The Agricultural Revolution

THE ESSENTIAL features of the agricultural revolution have been referred to in Chapters 2 and 3, which are concerned with the problems of the residues arising from large-scale applications of chemicals in farming. Modern developments in agriculture, which are still continuing, have taken place in a very brief period of time, which is probably no more than about 1 % of the total time in which agriculture has been practised. The main changes have involved the rapid intensification of methods of crop cultivation and animal stock rearing. The rapid changeover to homogeneous systems of mono-cultivation of crops specially selected for their high yield quality in a small number of larger-sized fields, coupled with the generous application of chemicals, poses the more immediate threat to both the external and internal environments. This situation is viewed with some concern by a growing number of scientists and has been well expressed by Dr Walter Ryder, formerly Head of the Entomology Department in the Institute of Animal Sciences of Havana: "Over the past few thousand years, man has developed a capacity now enlarging with unprecedented speed, to disrupt the processes of gradual biological evolution and ecosystem modification . . . It would seem logical to deal with the inevitable consequences (among them the problems of pests and diseases) on the principle of restoring to the environment some measure of stabilising heterogeneity."[3]

The green revolution involving the introduction of high yield varieties of cereal grains in a number of Asian countries has already made a notable contribution towards relieving those countries from near famine conditions and from their former dependence on imported grains. The latest annual review of the United Nations FAO, however, seems to imply that the initial rapid increase in crop production with the introduction of these special varieties is now slowing down. The 1971 rate of increase in production has been estimated at only about 1 % against an increase of about 4 % in previous years; the food output per person in these countries has actually declined owing to the more rapid increase in the population growth.[4] There are also fears that the new high yield varieties may have a lowered resistance to pest infestation and may require the continuing application of chemicals on a large scale for some time, or until such times as resistant strains are available and established.

The agricultural revolution has, therefore, been a major factor in introducing a wide range of pesticide chemicals into the human environment and the great majority of them in a very short interval of time. To these must be added residues of antibiotics and hormones used in intensive stock rearing.

HISTORICAL PERSPECTIVES IN INTERNAL CONTAMINATION

DURING THE whole of the period of man's gradual change-over from a nomadic to a settled existence dietary habits would also have been evolving gradually to satisfy essential human needs without any systematic knowledge of nutritional requirements. The ability of man to evolve a satisfactory diet even without the advantages of nutritional science has been demonstrated by the discovery in quite recent times of quite primitive and remote communities who have evolved for themselves a perfectly adequate diet based on a selection of locally available foods. To arrive at such a balanced diet in these circumstances probably involves the use of a long period of trial and error in which those foods which are beneficial are selected preferentially and those which are harmful are eliminated. Dietary habits which have evolved slowly to meet the instinctive needs of the people have only begun to change rapidly in very recent times. Harmful factors in the natural diets may not be entirely absent, but there would be an adequate time for the organism to adapt to any adverse factors.

The recent industrial changes referred to in the previous section have all been accompanied by the rapid urbanisation of the population. An immediate consequence is that more and more people are separated from the resources of the land and less dependent on their own methods of food cultivation and preparation. These changes are illustrated by the rapid expansion of bakeries at the expense of home baking at the early stages of the Industrial Revolution and also by the rapid development in the scale of mechanisation of food industries, such as biscuits, jams and confectionery. The development of bakeries was also accompanied by a switch from barley, rye and oats as the basic grain ingredients of the bread to wheat and a growing preference for white over brown bread. A similar trend was also observed in the growth of the breweries and the almost complete disappearance of home brewing. The coming of the railways helped to expedite the entry of fresh farm produce into towns and cities, but increasing reliance had to be placed on imports as the population expanded. There was, for instance, a considerable expansion in the amount of sugar consumed after about 1850. The most rapid changes in the average dietary habits have, however, been occurring since 1945 and accompanying the rapid changes taking place in the external environment. The most recent trends have involved the supply of an increasing number of refined and processed foods in cans or in packages, all designed to ease the tasks of food preparation in the kitchen. However desirable these changes may be, they place the consumer at one further remove from the original sources of food. The refining and processing of these foods demands a greater uniformity in the raw food materials, sometimes at the expense of quality, and frequently requires the introduction of the chemical additives to improve flavouring, palatability, texture, and shelf-life. Gross adulteration of food was not uncommon in the last century and the Food and Drugs Act of 1860 was introduced to exercise legal controls over the quality of the food offered to the public. The gross abuses of the 19th century are unlikely to be repeated; the problems have just become more complex and difficult to evaluate because of the large number of food additives which might be used at low concentrations.

The advantages of processing are essentially that necessary foods can be made available all the year round and that their nutritional quality and purity can, at least, be subject to measures of control. The disadvantages which have to be considered are that changes in the nature of the foods as a result of processing and the extent of the use of chemical additives may yet have long-term adverse effects on human health. The essential question relates to the ability of the human organism and especially the digestive system, which is the product of a long evolutionary development, to adapt entirely to the extremely rapid changes to which it is now being exposed.

The technological revolution in food processing and marketing in recent years and which is still gathering momentum, has taken place simultaneously with the other developments in agriculture, industry and medicine. The many chemicals which are present in food and in the body from these sources are additional to very many other substances which are present quite naturally. All natural foods are in fact quite complex mixtures of chemical substances including the major nutrients such as the carbohydrates, fats, proteins, mineral elements and other essential substances such as the vitamins and trace elements, natural colouring, flavour and aroma compounds present in small amounts. The chemicals which are used as food additives or are present as impurities, cannot be assumed to be automatically toxic to man and the essential problem is one of identifying those constituents, whether natural or artificial, which present long-term risks to the future health of the human population.

The many changes in food habits and the existence of food contamination are largely accomplished facts and may be very difficult to put into reverse. The human organism is extremely adaptable in coping with changes but it could certainly be argued that it is now being subjected to too many basic changes in too short a time. There are already sufficient warning signs that all may not be so well with the state of human health as we are sometimes led to believe. It is indisputable that developments in medicine and in public hygiene have extended the lifespan of the average individual, and have been particularly effective in reducing the incidence of infantile mortality. The annual death rates from infectious diseases, such as influenza, diphtheria, and tuberculosis, the major killers until recent times, have shown a progressive decline due to medical advances. Tuberculosis, for example, declined in the USA from about 2000/million of the population in 1900 to about 200/million in about 50 years. Despite this achievement, however, the death rates due to chronic causes, such as cardiovascular disease and cancers have shown a steady increase. In the USA, for example, over the same period of 50 years, the death rate due to all forms of cancer almost doubled from about 700/million of the population in 1900 and cardiovascular disease rose from 3400 to 5200/million population. In the case of cancers, there has also been a marked increase in the incidence of lung and

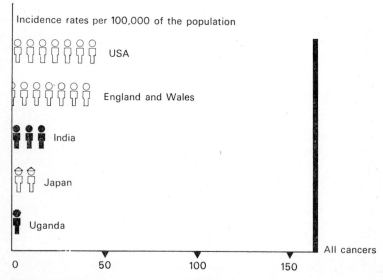

Fig. 7.1 Civilisation disease: cancer of the colon

colonary cancers which are now responsible for the majority of cancer deaths. The incidence of lung cancers is almost certainly due to non-dietary factors but the increase in colonary cancer may well be associated with changes in food habits. Recent epidemiological surveys into the incidence of colonary cancers in a number of countries including several of the developed countries with western standards of life and a number of the developing countries of the world suggest strongly that dietary habits are involved. The work of Dr Burkitt of the Medical Research Council has special relevance in this connection.[5] A selection of his results embracing the highest and the lowest rates of colonary cancer incidence are presented in Fig. 7.1. The high rates in the Western nations, which approach 500/million population per year, should be compared with the low rates in the developing nations, typified by Uganda, at about 50/million of the population per year. This marked difference may be associated with a preponderance of refined and pro-cessed foods in the diet of the Western nations compared with almost unrefined raw foods in the more primitive village communities of the developing nations. The role of the dietary factors was examined further by Burkitt by means of experiments involving small groups of children in an English boarding school, an African boarding school, and a Ugandan village. In each of these three small groups, the transit times for the passage of food through the alimentary tract were measured by incorporating small, totally inert, plastic pellets into the food. The transit times were also compared with the weights of faeces eliminated daily; all these results are presented in Fig. 7.2. They show striking differences between the groups, with the English boarding school-children having an average transit time of 90 hours and a small daily weight of faeces of just over 100 g, and the primitive community having a transit time of about 35 hours and an average weight of faeces of about 460 g per day. The African boarding school represents an inter-mediate stage of dietary development. These differences indicate strongly that dietary factors are having a marked influence on the digestive processes and one of the factors which might be involved is the amount of indigestible fibre or roughage in the diet. When these results are related to the epidemiological evidence of differences in the incidence of colonary cancers, they imply quite strongly that dietary factors have a dominant causative role. The precise factors and mechan-isms which are the direct cause of the cancer cannot yet be identified with certainty. The influence of roughage in speeding up the passage of food through the g.i. tract will certainly reduce the ʟesidence time of the non-absorbed portions of the diet in the lower digestive tract and in the large intestine, limiting the time available for any interactions between contaminants and the surfaces of sensitive tissues. The reduced contact time could, therefore, have considerable advantages to the

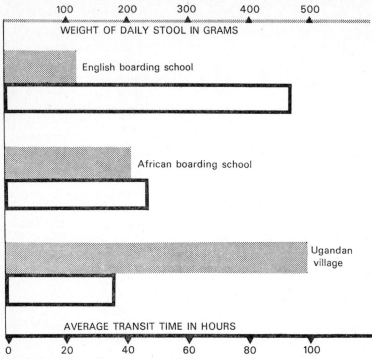

Fig. 7.2 Digestion of food: the transit times for the passage of food through the alimentary tract in three different groups

organism. As many agents are eliminated in the faeces, either by failure to be absorbed during their passage through the small intestine or after re-entry via the entero-hepatic cycle after being absorbed, this possibility cannot be discounted. A further possibility affecting the whole body is that the more rapid transit times will also result in a smaller uptake of any contaminating agent through the small intestine. Birkitt himself, however, considers that changes in the populations and strains of the colonies of micro-organisms inhabiting the large intestine, and the lower part of the small intestine may be more directly responsible. Although this cannot be confirmed at the present moment, it could have special significance when residues of antibiotic are present in food or when they are used in medicine.

The importance of Burkitt's work is the clear demonstration that dietary habits have a marked influence on the digestive processes in man; this may also subsequently influence the elimination and distribution of the contaminating agents within the body. Other dietary factors, in addition to the content of indigestible fibres, may also contribute to the problems of chemical contaminants and the health of the individual. One of these factors might be the considerable increase in the quantities of refined sugars and starches consumed in the average Western diet. The average sugar consumption in England has increased almost twentyfold in the last hundred years and Yudkin has presented evidence relating this increase to the growing incidence of cardiovascular disease.[6] Cleave, Campbell and Painter[7] have also reviewed the evidence for refined sugars and starches as contributory factors to cardiovascular diseases and the other civilisation diseases in Western countries. There seems little doubt, therefore, that one or more of these factors associated with changes in food habits and the composition of the foods consumed may well be associated with the rising incidence of chronic illness in the West. They lend weight to the central thesis of Dr R. J. Williams "that the nutritional micro-environment of our body cells is crucially important to our health and that deficiencies in this environment constitute a major cause of disease".[1] Williams is, of course, concerned primarily with the deficiencies in the nutritional quality of the diet but if these deficiencies are extended to include the various food contaminants, they may well reinforce his basic premise.

TESTING FOR TOXICITY

THE CONSEQUENCES of tissue exposure to the variety of contaminants or to their metabolic products at low concentrations after passage through the alimentary tract remain the central problems. Any estimates of the results of such exposures are almost exclusively based on studies involving a single agent and, more occasionally, involving pairs of small groups of agents. The great majority of the work which has been reviewed in the earlier chapters appears to reach at least a measure of agreement that the available evidence for enabling sound judgements in such a complex situation is either quite inadequate or just not available. "At the moment, we know practically nothing of the long-term effects on the body of an ever-increasing number of food additives and non-essential medications, and yet we are all exposed to this hazard which could change our physiology and change our disease patterns if allowed to continue unrecognised . . . Our ignorance is abysmal; even more frightening is our lack of regard for the problem." Professor Beaconsfield's concern is only reinforced by the presence of all the environmental pollutants.[8]

The problems of testing, with special reference to drugs and food additives, have been discussed in some detail in the paper by Professor Beaconsfield. It is, of course, well known that drugs after tests on animals are usually subject to clinical trials on closely observed human beings, but it is also apparent that it is impossible to design trials which are completely foolproof. The prediction of human toxicity from animal experiments is also subject to very serious limitations and especially when different reactions and metabolic patterns between species are to be expected. Professor Beaconsfield suggests that it is quite possible "to obtain a so-called normal physiological picture which would not reflect underlying physico-biochemical alterations", which could lead ultimately to irreversible damage, or side-effects in the exposed organs. These are a manifestation of disturbed biochemical and biophysical processes and it is imperative, therefore, that there should be a much closer understanding of the precise mechanisms of damage and that these should be studied at the tissue and cellular levels. "Such testing should tell us whether the cells of the body's vital organs are deranged functionally or architecturally by exposure to the compounds; whether any changes which do occur are cumulative and reversible on withdrawal of the compounds; and whether any cytogenic effects are demonstrable."[8] It is recognised that such tests present a formidable undertaking and that there is an urgent need for scientists of multidisciplinary training to undertake the work. Wherever possible, experimental evidence of toxic effects should also be backed up by epidemiological studies on small groups of a population selected for their known exposure to higher levels of the contaminant. The special studies which the World Health Organisation are undertaking at a few places which have higher than normal natural radiation backgrounds, such as at Kerala in India and at La Paz in Bolivia, are a good example of this type of investigation.

The limitations in predicting human toxicity from animal experiments are due mainly to variations in the response of the different species and their varying sensitivities to different agents. These variations may be due to differences in the metabolic pathways and to differences in tissue sensitivity to the agent or its metabolites. There is also the experimental problem of having to use large numbers of animals in order to obtain statistically relevant data. As an example, it may be necessary to use as many as 30,000 animals to demonstrate the incidence of an effect such as cancer which might occur in man at a rate of 1 in 10,000 from exposure to contaminated food. The only practical alternative to using large numbers of animals is to increase the level of exposure well above the expected levels in food. This procedure then involves a dubious double extrapolation from high to low exposure in the animal experiments and the further extension of the results to man himself. Studies of the genetic effects from exposure to ionising radiation provide an example of one of the very few large-scale experiments involving large numbers of animals. Even in this case there was still an appreciable gap between the lowest experimental exposures and the exposures due to food contamination.

What is therefore required at the moment is a comprehensive examination of the effects of the many agents already present in food, not only to extend the studies in progress on individual agents, but to develop the work to include groups of agents, and especially those groups where there are grounds to suspect that interactions may be occurring. The formidable nature of such a programme involving the present techniques of experiments on animals and their doubtful significance in terms of human effects requires a fresh approach to the whole problem. Although the alternative methods involving tissue cultures may also have their limitations when the effects have to be extrapolated to the whole organism they should be capable of revealing the mechanisms of damage and the nature of any interactions which are taking place at the cellular level. It may then become possible to make reasonably confident predictions of the consequences of human exposure and especially to screen the consequences of any new agent entering the environment. The task is however so formidable that to stand any chance of success it must command the support of governments and international agencies.

Any agent of pollution in addition to satisfying the first criterion of safety, assessed on the best available evidence, should also satisfy two other criteria of efficacy and identity.[9] The criterion of efficacy requires that any benefits in using a chemical agent must be sufficient to far outweigh any potential hazards. As there is an obvious element of risk in most human activities (*see* Chapter 1, p. 33), this striking of a balance between the benefits and the risks calls for careful and considered judgement, and should also have the acquiescence of the general public. This is obviously a difficult matter, but the difficulties should not be allowed to stand in the way of a properly informed public. The third criterion, that of identity calls for a clearly defined specification of the composition and purity of any chemical agent, perhaps on the same lines as the WHO International Pharmacopeia of drugs and medicines. This criterion is necessitated by virtue of the processes used in the manufacture of the chemical, and the probable presence of impurities which even in very small concentration may be more toxic than the principal agent. The presence of trace amounts of dioxin in 2, 4, 5-T illustrates this particular problem.

REMEDIAL MEASURES

ANY IMPROVEMENTS in the quality of the external environment should be very helpful in the course of time in reducing the risks of food contamination. These improvements will concern the manufacturing industries and especially those involved in the production of chemicals and the generation of power at the lowest practical cost. The products and wastes of these industries are accidental food contaminants and are referred to on p. 223. Modern agricultural practices are another source of food contamination and the problems of reducing this source are referred to below:

(a) Measures to control agricultural pollution

RESIDUES OF crop protection chemicals in food are a major source of present day human pollution. In addition to their threat to human health, there is also an increasing concern regarding the efficacy of their applications in agriculture and especially the insecticide chemicals. This concern arises for two different reasons:

1. The indiscriminate nature of the attacks of these chemicals on insects, both those which are pests and others which may be beneficial. The damage to honey bee colonies reported several years ago due to the persistent chlorinated hydrocarbons is one such example.
2. The emergence of certain insect pests with an acquired resistance to the effects of the chemical. When the insect pest is able to acquire this resistance sometimes more readily than other insects which may help to keep it under control, the pest may even become more firmly established than it was prior to treatment. There are now many examples of the emerging chemical resistance, one outstanding example being the malaria-carrying mosquito. For all these reasons, various alternative systems of protection such as more specific chemicals or biological controls are the subject of a great deal of present attention.

An effective long-term alternative to the present pesticide chemicals in the context of modern agricultural practice is not yet available and many of those which are under investigation may only be temporary palliatives. It is, therefore, reasonable and it is becoming increasingly practical to consider a genuine and perhaps very radical alternative approach which would not only relieve the worst aspects of crop pollution but would, at the same time, make a very positive contribution towards solving the intransigent world food problem which remains a matter of some urgency. Both of these problems could be effectively tackled by the direct conversion of agricultural crops, such as several varieties of bean and cereal into acceptable analogues of meat and milk, by-passing in the process the highly inefficient animal. The majority of the world's population already relies very heavily on plant sources for its essential protein as is illustrated in Table 7.2. The dependence of the world population directly on plant sources for body energy is even more pronounced. The P/A ratio for protein and the whole world

Table 7.2

The sources of protein in 1960[10]

| Source | Protein consumption, Mtonnes/yr. | | |
	Developing countries	Developed countries	Whole world
Animal produce (A)	6·6	14·8	21·4
Plant produce (P)	34·7	12·6	47·3
P + A	41·3	27·4	68·7
P/A	6·6	0·85	2·2
Population	2·5G	1·00G	3·5G

It has also to be noted that the 5/7ths of the total population living in the developing nations has only about half of the total land available for cultivation and this is generally less fertile than the available land in the western nations. The average income of the population in the developing nations is only about 1/10th that of the western nations so that any attempt to follow western agricultural practices with emphasis on animal proteins can only be disastrous for the developing nations
N.B. $IG = 10^9 = 1000$ million.

population is increased to 9·0 if a further 135 Mtonnes of protein for feeding livestock is included in the total. This is mainly because the average protein conversion efficiency of farm animals is extremely small and usually less than 10% (the average protein conversion efficiency is the percentage of edible protein produced by the animal over the protein which it receives in its feedstuffs). It is the huge demand for animal feedstuffs which is responsible for most of the enormous pressure on the land to increase its productivity. The various processes of converting crops into acceptable protein foods which are now readily available would operate at much higher efficiencies approaching 100%. This would relieve the pressure on the land very substantially and the necessity to force increasing production off it to satisfy the rising demand. The necessity to resort to intensive stock rearing would also be reduced and the risks from pathogenic organisms, antibiotic and hormone residues would also be drastically decreased. The analogues of meat and milk which can now be prepared from plant sources can be made perfectly acceptable and nutritionally adequate with the further advantage that the quality of the product can be very carefully controlled. A number of these products are already available on the market. The additional bonus is that food crops, originating sometimes in the developing countries, can be diverted immediately to their own urgent needs.

An immediate and further consequence of relieving the pressure on the land would be the opportunity to resume methods of crop cultivation based on more traditional systems of diversity and rotation. This is important in contributing to Ryder's principle of restoring to the environment some measure of "stabilising heterogeneity", and thereby to ensure that "in carving uniformity out of heterogeneous ecosystems" the point is not reached, "where the quality of life is bound to suffer"[3]. This can be achieved by the rotation of crops or by the deliberate interplanting of crops in the same field. A reduction of 50% in the incidence of insect pests in maize was attained recently in Cuba by sowing sunflowers in eight-metre strips in between the sowings of the maize. Heterogeneity in crop production may also encourage the introduction of some of the other alternative and less drastic methods of controlling the worst ravages of pests such as the biological control systems. These can assume a variety of forms, a classical example being the introduction of an Australian ladybird into the California orange groves to control the attacks of the citrus scale beetle which were quite serious at the beginning of the century. This proved highly successful until the fruit growers decided to introduce DDT about 1945; this proved to be more harmful to the predatory ladybird than to the beetle pest, which was enabled to re-establish itself in greater intensity than ever before. This example of insect control by a biological method is also matched by another example of plant weed control; this involved the introduction of an Argentinian moth into the Australian outbacks in 1930 to control the spread of the "prickly pear" and to preserve land for grazing purposes. Other systems of biological control which are being developed and in some cases tried out in the field, include viral and bacterial controls, and the release of male insects sterilised by an irradiation treatment in the pupal stage.

There may also be a need to continue with the application of certain chemicals subject to stringent regulations and safeguards. These could contribute to a state of heterogeneity in the insect population of the environment, and therefore play an effective part in an overall pattern of pest control. They could include some of the chemicals in use at the moment but possibly treated with protective coatings to prolong their lifetime in the field and ensuring that they become available only at the appropriate time; these could be supplemented by others of a more specific nature for the insect hosts which they attack. They could also include minute amounts of other special chemicals which are able to interfere with the weak chemical signals which direct the behavioural responses of the insects. These could be used to mislead an insect in the selection of the host plant for its food or for the deposition of its eggs; they can also be used to affect the sexual behaviour of the male insects.

The problems associated with the indiscriminate use of pesticide chemicals are now much better appreciated, and there has been an encouraging response from various governments in banning or in restricting the applications of some of the more persistent chemicals,

especially aldrin, dieldrin and DDT. Another form of agricultural
pollution, involving the contamination of animal produce by pathogenic
strains of bacteria with acquired resistance factors to certain antibiotics
used in human medicine, is also expected to show some improvement
in the UK; this follows the government's decision to make it unlawful
to sell feedstuffs containing certain antibiotics, such as penicillin and
the tetracyclines, used in human medicine.

(b) Measures to control industrial pollution

INDUSTRIAL CONTROL of pollution is very largely a matter of economics
and could be achieved in very large measure by increased expenditure
on the treatment of the various forms of wastes using technical processes
which are already available. The control of industrial discharges into the
environment is therefore very largely a matter of prevention by adequate treatment at source. Some measures of control over industrial
pollution and of preventing any substantial increases could undoubtedly
be achieved by curbing the rate of economic growth and especially by
restricting the rate of growth in the manufacturing and power industries.
The manner in which the demand for power multiplies at a faster rate
than that of the growth in population suggests that there will be considerable difficulties in realising any effective curbs on economic growth
in the foreseeable future. The general public would clearly be reluctant
to sacrifice any of the material benefits which it enjoys at the moment,
and even if these were to be at the expense of reduced contamination of
the environment. It has also been made perfectly clear that any attempt
to restrict economic growth in the developing countries will be quite
unacceptable to the countries concerned. It would appear therefore
that the only effective controls will have to depend on technical measures
for processing the wastes of industry and especially for eliminating any
sources of toxicity in the wastes. This will increase the costs of production inevitably, and these will have to be borne by the public either
through increased costs for the products of industry or through government grants with the costs shared by all members of the public. There
is certainly no technological reason why the worse effects of industrial
pollution such as waste effluents into rivers and coastal waters should
not be brought rapidly under much greater control. The British government has recently announced vastly increased sums of money to clean
up the rivers, and the necessary legal powers such as the Clean Rivers
Acts of 1960 are also available. Another important measure is to restrict
the amount of lead which is permitted to be added to petrol; the first
steps towards this have also just been announced by the British government, although they may be expected to have only a marginal effect
initially.

The growing dependence of the power industries on nuclear energy sources presents a number of special risks. These include the rather unlikely but potentially serious type of accident to a nuclear reactor involving the loss of coolant in the reactor circuit leading to a possible failure of the containment systems. This problem has given rise to much concern in the USA in recent years in the case of the Light Water Reactors. This type of accident involving a reactor and, to a lesser degree, an accident during the transport to the chemical separation plant of highly radioactive fuel elements discharged from the reactor cores could lead to locally high contamination levels from escaping fission products. There are also the problems of coping with the tremendous quantities of fission product wastes after the chemical separation process, a problem which is dealt with in the UK by concentrating them into a small volume and storing as liquid waste. The quantities of such waste will grow as the nuclear power programme develops and storage will be required for at least a hundred years because of the considerable amounts of long-lived fission products such as strontium-90 and caesium-137 in the wastes. Alternative methods of storage are being investigated such as conversion into solid glassy materials in which the active materials would be effectively trapped. These problems will remain so long as the nuclear power programme is dependent on nuclear fission, but will be very largely eliminated if and when nuclear fusion is able to take over, perhaps towards the end of the century.

There is also mounting concern for the activities of small terrorist groups and the possible hijacking of radioactive materials and especially the fissile element plutonium produced in ever larger quantities as the nuclear power programme develops. Even if this plutonium could not be diverted to the production of atomic weapons, it might be used to pose a threat to water supplies. It is evident, therefore, that strict security measures for the storage and transport of this material are absolutely essential.

(c) Further requirements for effective pollution control

THE EXISTENCE of legislation is in itself insufficient to control excessive pollution unless it is supported by powers of inspection and is backed up by the resolve of an informed public. Many of the legal powers to control the worst excesses of pollution in the UK have been available for some time; they may have been responsible for avoiding even worse pollution of rivers and estuaries but they did not succeed in preventing many of the abuses which exist today.

The effectiveness of any measures which are taken to control pollution at its source will still require enforcement through adequate systems of monitoring and inspection. The UN at the recent Stockholm Conference on the Environment agreed to the setting up of global systems of monitoring for certain pollutants, and these are almost certain to be backed up progressively by national arrangements. The regular monitoring of foods for radioactive fall-out and certain pesticide residues is well established in Britain, even though there has been a considerable reduction in the number of food samples checked regularly for radioactive fall-out. A scheme for the monitoring of some of the heavy metals in fish and other foods which are important in the national dietary has also been put into operation. All such measures are to be welcomed and encourage the hope that increasing information can help to pinpoint the major sources of pollution and provide invaluable assistance in planning the necessary control measures.

Many other suggestions have been made from time to time to improve the effectiveness of any steps to control pollution, and many of these are reviewed in the paper by Epstein.[9] They include such measures as:

1. The creation of international data banks or a registry of all chemicals which may pose problems of pollution; the chemical specifications of such compounds could also be included, in addition to what is known about their efficacy, toxicology, results of monitoring, and epidemiology. It might also be possible to integrate any such arrangement with an international regulatory and early warning alert system.
2. The enforcement of existing national legislation or the enactment of new legislation to impose the required standards of control, to ensure that all issues relating to human safety and environment quality, and all data relevant to such discussions, are made public, and finally to ensure full measures for impartial and competent testing of the sources.
3. The establishment of multi-disciplinary training courses and research programmes into the whole range of environmental problems.

The success of all attempts to improve the human environment, externally and internally, will depend ultimately on the will of the population expressed through the media and national governments. There can be very little doubt about the growing anxiety of the people, which it would be unwise to ignore. There can be no doubt whatever that there is also a serious need for much more information, to enable the whole range of problems to be analysed and discussed in a cool and rational manner, and not in the fragmentary fashion which has largely prevailed up till now. In the words of John T. Edsall of the Biological Laboratories of Harvard University, "I am not one of those who is crying out 'Doom within the Decade', but as far as keeping the world a livable place is concerned, I regard present trends as extremely ominous. As one example I suspect a survey of the world would show a steady increase over the last half century and more, in the amount of land that is becoming desert or semi-desert . . . I suspect that modern technology, and the need of feeding a rapidly rising population, are accelerating the process. I want to see mankind develop a world that will be at least as good to live in a hundred years hence, as it is today. I am not giving way to despair and I believe in working to change what I see as the present trends; but I cannot be an optimist."[11]

This concern is also shared by the President of the US. In sending the first annual report of his Council on Environmental Quality to Congress, President Nixon stated that "our environmental problems are very serious, indeed urgent, but they do not justify either panic or hysteria. The problems are highly complex, and their resolution will require rational, systematic approaches, hard work and patience." There is no point in attempting to conceal the problems and there must be an open and frank discussion of all the possible consequences and not least the problems of the human internal environment, which may affect us all in some degree, and may have even more serious consequences for the health of future generations, if they are not evaluated with a full sense of urgency.

REFERENCES

[1] *Nutrition against Disease: environmental protection*, Roger J. Williams, Pitman, London, 1971.

[2] *A History of Europe*, H. A. L. Fisher, Edward Arnold, London, 1949, p. 11.

[3] "Agriculture: the roots of deterioration", Walter Ryder, *New Scientist*, 1972, **54** (8th June), 567.

[4] Annual Review of the UN Food and Agricultural Organisation (FAO) reported in the *Financial Times*, 21 November 1972.

[5] "Related Disease-Related Cause", D. P. Burkitt, *Lancet*, 1969, 1229, and *Cancer*, 1971, **18** (1), 3.

[6] "Sugar and Disease", John Yudkin, *Nature*, 1972, **239**, 197.

[7] *Diabetes, Coronary Thrombosis and the Saccharine Disease*, T. L. Cleave, G. D. Campbell, and N. S. Painter, John Wright, Bristol, 1969.

[8] "Internal Pollution: our first priority", Peter Beaconsfield, *New Scientist*, 1971 (18th March), 600.

[9] "Control of Chemical Pollutants", Samuel S. Epstein, *Nature*, 1970, **228**, 816.

[10] "The Role of Plant Foods in Solving the World Food Problem": proteins", F. Wokes, *Plant Foods for Human Nutrition*, 1968, **1** (1), 23.
[11] "Doomsday Syndrome", John T. Edsall, *Nature*, 1971, **234**, 56.

Suggestions for further reading
Toxic Constituents of Plant Foodstuffs, ed. Irvin E. Liener, Academic Press, New York and London, 1960. (A useful survey of natural toxic constituents of plants, with a chapter on adventitious toxic factors in processed foods.)
Plenty and Want: a social history of diet in England from 1815 to the present day, John Burnett, Penguin, Harmondsworth, 1968. (Describes the changing patterns of food habits during the industrial revolution.)

Appendix

THE PERIODIC TABLE OF THE ELEMENTS

GROUP PERIOD	I	II	3	4	5	6	7	8			1	2	III	IV	V	VI	VII	0
1	1 H																	2 He
2	3 Li	4 Be											5 B	6 C	7 N	8 O	9 F	10 Ne
3	11 Na	12 Mg											13 Al	14 Si	15 P	16 S	17 Cl	18 A
4	19 K	20 Ca	21 Sc	22 Ti	23 V	24 Cr	25 Mn	26 Fe	27 Co	28 Ni	29 Cu	30 Zn	31 Ga	32 Ge	33 As	34 Se	35 Br	36 Kr
5	37 Rb	38 Sr	39 Y	40 Zr	41 Nb	42 Mo	43 Tc	44 Ru	45 Rh	46 Pd	47 Ag	48 Cd	49 In	50 Sn	51 Sb	52 Te	53 I	54 Xe
6	55 Cs	56 Ba	57* La	72 Hf	73 Ta	74 W	75 Re	76 Os	77 Ir	78 Pt	79 Au	80 Hg	81 Tl	82 Pb	83 Bi	84 Po	85 At	86 Rn
7	87 Fr	88 Ra	89** Ac															

*	LANTHANONS RARE EARTHS	58 Ce	59 Pr	60 Nd	61 Pm	62 Sm	63 Eu	64 Gd	65 Tb	66 Dy	67 Ho	68 Er	69 Tm	70 Yb	71 Lu

**	ACTINONS	90 Th	91 Pa	92 U	93 Np	94 Pu	95 Am	96 Cm	97 Bk	98 Cf					

Index

PRINTED IN GREAT BRITAIN BY
COX & WYMAN LTD
LONDON, FAKENHAM, AND READING

fish remaining a major dietary source of mercury, and especially organo-mercurial compounds, several countries have taken steps to apply limits to the consumption of contaminated fish. As an example, the Swedish government has imposed a total ban on the consumption of fish when the mercury levels exceed 1 mg/kg with a further recommendation that fish with concentrations between 0·2 and 1·0 mg/kg should only be consumed once per week. In Canada and the USA, a ban on fish consumption has been imposed when the concentrations exceed 0·5 mg/kg. The problems of mercury contamination have also been considered by a Joint Party of FAO and WHO and an upper limit for the total concentration in food consumed of 0·05 mg/kg (50 ppb) has been recommended. This would be equivalent to a maximum daily consumption of 80 μg Hg. The average intake from food consumed in the United Kingdom, about 10 μg per day at a concentration of 6 μg/kg, is therefore well below the recommendation of the Joint Committee and should give rise to little cause for anxiety at the present moment, with the possible exception of the small groups of the population depending on the higher intake of fish.

ARSENIC

NON-METALLIC ELEMENTS as food contaminants have been mentioned briefly in the introduction to this chapter and this section is concerned with one of the better known of these substances. The toxic effects of arsenic have been well known for a long time but it is, perhaps, not quite so readily appreciated that it occurs at very low concentrations in all plant and animal tissues, although its functions are totally unknown. Its average concentration in human hair is about 1 ppm. Outbreaks of food poisoning due to arsenic have occurred several times in the past in some cases as a result of its presence in sulphuric acid, which is used in certain food processing plants. Arsenic was also much used in the past in certain compounds as a pigment and, as such, has been suspected of being a contributory factor to the death of Napoleon. More recently, a number of inorganic compounds of arsenic, for example lead arsenate, have been used as insecticide sprays in orchards, and a few organic compounds of arsenic have been developed for use as contact herbicides. A further possible source of food contamination is the drug arsanilic acid, which was previously used in the treatment of certain diseases, and which is also added to animal feeds as a growth promoter. Its earlier use in medicine was discontinued owing to its tendency to produce blindness by atrophying the optic nerve. The normal soil concentration of arsenic is about 5 ppm.

The responsibility for the analysis of arsenic in foods for human consumption in the UK has been undertaken by the Joint Committee for Pesticide Residues. The results reported by the Committee are mainly for total arsenic and make no distinction between the organic and inorganic forms of the element. The higher concentrations of arsenic are found in certain sea foods and indicate quite high concentration factors from the sea-water (Table 5.15). Although these represent

Table 5.15

Concentrations of arsenic in sea foods[12]

Sea food	Concentration ppm	Concentration factor = $\dfrac{\text{Concentration in food}}{\text{Concentration in sea water (3 ppb)}}$
Oysters	3–10	1000–3300
Lobsters	70	2300
Mussels	120	40000
Prawns	170	57000
Fish	<0·3	100

quite high concentrations, they do not contribute all that significantly to human contamination in view of the low average consumption, about 1·5%, of these foods in the normal diet. A diet based predominantly on sea foods, however, could obviously give rise to significantly higher arsenic levels in the body.

Table 5.16

An estimate of daily dietary intake of arsenic, 1967–8 based on the Joint Survey of Pesticide Residues in Foodstuffs sold in England and Wales[12]

Food Group	Consumption % w/w	Estimated concentration ppb Maximum	Average	Consumption µg/day Average
Cereals	16	310	175	44
Meat, fish	13·5	600	170	36
Fats	3	670	345	17
Fruits	13	2800	190	38
Root veg.	17	590	265	69
Green veg.	7	630	300	30
Milk, cheese	29	690	220	99
Misc.	1·5	–	–	–
Totals	100 (1·55 kg/d)			333

The normal average human diet is estimated to contain about 0·3 mg arsenic/day at an average concentration of about 0·2 ppm. This intake may be compared with the values at which toxic symptoms may appear, about 5 mg/day at a concentration of about 3 ppm. The quantity and concentration which may lead to fatal poisoning is about 100 mg/day or 60 ppm in the diet, respectively. The sources of dietary intake are presented in Table 5.16.

The calculations of the dietary consumption, which are based on the results of the Joint Committees, suggest that the daily intake averaging just less than 0·35 mg/day could rise to a maximum of about 1·5 mg in extreme cases. It is not possible to say whether the average intake of this magnitude is really higher than the average intake prior to the agricultural applications of arsenic as a pesticide, or its use as a growth promoter. Any effects of such a small increase would, in any case, be very difficult to assess. It is also possible that any slight increase due to agricultural practice is offset in this case by some recent reductions in industrial discharges of the element.

The arsenic is widely distributed in body tissues being present to the extent of:

0·5   ppm   blood
0·1   ppm   liver
0·01 ppm   bone (where it may replace phosphorus partially as phosphate)

Only in the case of the blood does there appear to be a slight concentration effect from food to man.

It is extremely unlikely that there would be any chronic effects due to arsenic exposures at the present concentrations as these only begin to appear above 3 ppm and usually involve non-specific symptoms such as a general weakness, malaise, pains in the stomach and limb extremities. Arsenic is, however, suspected of being a potential carcinogen with the liver as the most probable site in the case of arsenic ingested from food.

One possible beneficial effect of arsenic appears to be a measure of protection it can afford against excessive amounts of another non-metallic substance, selenium. The latter is an essential element at low concentrations and toxic at concentrations not many times greater than the normal safe requirement. The situation may have some similarities with that existing between copper and zinc, with arsenic and selenium being in competition in certain physiological functions.

REFERENCES

[1] M. R. Spivey Fox, in *Metallic Contaminants and Human Health*, ed Douglas H K. Lee, Academic Press, New York and London, 1972, p. 191. (The principal chapters are devoted to Cd, Hg, and Pb, with smaller contributions on Be, Cr, Mn, Ni, V, As and fluorides.)

[2] (a) *Trace Elements in Biochemistry*, H. J. M. Bowen, Academic Press, New York and London, 1966 (contains a wealth of information about all naturally occurring elements and their biospheric distribution).
(b) Recommendations of the International Commission on Radiation Protection, Report of Committee II on Permissible Dose for Internal Radiation, ICRP Publication 2, Pergamon Press, Oxford, 1959.

[3] "Heavy Metal Concentrations in Coastal Waters", M. I. Abdullah, *Nature*, 1972, **235**, 158. (See also the Royal Commission 3rd Report on Environmental Pollution, *Pollution in some British Estuaries and Coastal Waters*, Cmnd 5054, HMSO, London, 1972.)

[4] "Essential Trace Metals in Man: Zn, relation to environmental cadmium", H. A. Schroeder, *J. Chronic Diseases*, 1967, **30**, 179–210.

[5] "Aspects on the Toxicity of Cadmium and its Compounds", Nilsson, R., *Ecological Research Bulletin No. 1*, Swedish National Research Council, Stockholm.

[6] "Copper and Zinc Status of Ewes and Lambs Receiving Increased Dietary Concentrations of Cadmium", C. F. Mills and A. C. Dalgarno, *Nature*, 1972, **239**, 171.

[7] *The Limits to Growth*, D. H. Meadows *et al.*, a Potomac Associates Book, Earth Island Ltd., London, 1971. (The classical study by the Club of Rome of the basic problems of population growth, resource exploitation, and pollution.)

[8] "Mercury in British Fish", G. Grimstone, *Chemistry in Britain*, 1972, **8**, 244.

[9] *Environmental Mercury Research in Sweden*, J. E. Larsson, Report of the Swedish Environmental Protection Board, Research Secretariat, Stockholm, June, 1970.

[10] "Residual Mercury Content of Seed Potatoes Treated with Organomercury Disinfectant", G. A. Hamilton and A. D. Ruthven, *J. Sci. Fd. Agric.*, 1967, **18**, 558.

[11] (a) *Survey of Mercury in Food*, 1st Report of the Working Party on the Monitoring of Foodstuffs for Mercury and Other Heavy Metals, HMSO, London, 1971.
(b) Report of the Government Chemist, 1971, Laboratory of the Government Chemist, Department of Trade and Industry, HMSO, London, 1972.

[12] Joint Survey of Pesticide Residues in Foodstuffs sold in England

and Wales, 1st August, 1967 to 31st July, 1968 (second year), published for the Association, by the Association of Public Analysts, London, 1971.

[13] "Mercury in the Environment", L. J. Goldwater, *Scientific American*, 1971, **224** (5), 15.

Suggestions for further reading
Environmental Mercury Contamination, ed. R. Hartung and B. D. Dunman, Ann Arbor Scientific Publications Institute, Michigan, 1972 (based on the proceedings of an International Conference on Mercury at the University of Michigan).

Health Hazards of the Human Environment, WHO, Geneva, 1972, chapters 3 and 12.

Addendum
The 16th Report of the Joint FAO/WHO Expert Committee on Food Additives (Technical Report Series No. 505, 1972), contains the following recommendations for "provisional tolerable weekly intakes" for the two elements:

1. *Cadmium.*
   400–500 µg per individual, equivalent to 57–71 µg per day, and a food concentration of 37–46 ppb. *N.B.* The Report also suggests that dietary intakes may vary from < 50 µg to 150 µg per day. See page 162.
2. *Mercury.*
   0·30 mg per person, equivalent to 43 µg per day, and a food concentration of about 30 ppb. (*N.B.* this supersedes the recommendation given on page 181).

6

Lead

INTRODUCTION

LEAD IS a further example of a moderately abundant, naturally occurring metallic element which has been closely and continuously associated with human activities since classical times. It was widely used by the Romans in the construction of a variety of domestic utensils, in some cases for the storage of wines to which it contributed an extra spice. As a result, quite high lead concentrations have been found in the bones of wealthy Roman citizens of the past and it has even been suggested that lead may have been a factor contributing to the downfall of the Roman Empire. In more recent times, lead has found extensive usage for piping water supplies and for the roofing of buildings, a variety of compounds of lead have been widely used as pigments in paints, as insecticides, and as an additive to petrol. Despite the great variety of applications involving lead and its compounds awareness of the possibility of the element becoming a general human contaminant, with harmful long term chronic effects has only been considered seriously in quite recent times, more especially since the development of the modern, high compression, car engine and the increasing use of lead as an additive to the petrol. A good example of the recent growth of interest in the chronic effects of lead exposure has been provided by the investigations of the New York City Health Department's Bureau of Lead Poisoning Control. This work which involved originally the screening of approximately 79,000 children revealed about 2,500 "positive" cases of lead concentrations in the children's blood. Altogether a total of 31,000 cases of children with "positive" levels of lead, including a possible 4500 cases of severe poisoning, had been detected by 1970; chipped paint in the deteriorating apartments of older tenement houses and the habit of pica (i.e. the habit of putting things in one's mouth) was identified as the prime source of this contamination. All the paints invariably included lead pigments, and notably white lead, as a base and this gave a sweetish taste to the flaking paint which made it attractive to young children. This practice can lead to high contamination levels, which may be far more serious for young children than later on in adult life.[1] Lead is also sometimes found as a filler in plastics, and plastic toys, or toys which have been coated with lead paints, have also been suspected as sources of lead contamination for children. The paint on British toys can quite legally contain up to 0·5% lead.

Paint chippings will also find their way into food during handling and preparation, providing an additional source of contamination for all members of the family. This source of food contamination may be further supplemented by concentrations of lead in drinking water supplies. This possibility is also greater in older houses where lead is used for piping the water supplies. These various sources of food contamination, although they can lead to intakes of lead from food or drinking water, will be of a more localised character than the general food contamination arising from lead which has been dispersed into the atmosphere from the car engine exhaust.

The development of the modern, high compression petrol engine, together with the rapid growth in the number of vehicles using the roads, has brought with it a train of human problems which are a penalty for the success with which the highly efficient modern engine has been developed. The high level of emissions from vehicle engines during the period of peak travel is a much-publicised problem with quite serious consequences in several major cities of the world, notably in the USA and Japan. The various gases which are discharged from the exhaust system include carbon monoxide, oxides of nitrogen, various partially burnt hydrocarbons together with lead. The mixture of gases when it is associated with local climatic factors can result in severe atmospheric pollution through the persistent photo-oxidant smog which is produced. This creates a health hazard to the local inhabitants inhaling the contaminated air, but two of the components of the discharge, namely the oxides of nitrogen and the lead, may create more widespread environmental and food pollution. The oxides of nitrogen are oxidised to nitric acid which when washed down by rain can result in a significant increase in the nitrate content of soils over and above the amounts which are added as mineral fertilisers in general agricultural practice (*see* Chapter 3). The lead compounds may similarly become widely dispersed, finding their way into soils and the produce of the land especially close to busy highways and in urban areas causing a general increase in the lead pollution levels.

LEAD IN THE ENVIRONMENT

LEAD IS a fairly abundant element in the earth's crust, having an average natural occurrence in the igneous rocks of the earth's crust of about 16 ppm. It is, therefore, present in all parts of the environment, including the biosphere and estimates of the distribution of lead in the various parts of the environment are summarised in Table 6.1. Several

Table 6.1

Lead in the natural environment
(ref[2a] Chapter 5)

Marine environment		Terrestrial environment	
		Earth's crust	16 ppm
Sea water	0·3 ppb	Soils	10·0 ppm
Plants	8·4 ppm	Plants	2·7 ppm
Fish	0·5 ppm	Animals	2·0 ppm

mineral sources of lead are also fairly widely distributed and include the principal ore galena (lead sulphide) with smaller amounts of cerussite (lead carbonate) and anglesite (lead sulphate). It has been estimated that about two million tons of lead are mined every year, of which about one-fifth is converted into the organic lead used as the petrol additive and most of which must find its way back eventually into the environment. The rest of the lead is employed either as the metallic element in a variety of alloys or is converted into the various compounds used for example as insecticides, and pigments. Many of the industrial operations involving the processing of lead result in some discharges of the element into the atmosphere and into surface waters, and it is possible therefore that rather more than one-fifth of the total lead mined in a year finds its way back into the environment. This quantity of lead returned to the environment, about 400,000 tons per annum, is approximately double the natural rate at which lead occurring in rocks

Table 6.2

Applications of lead which may cause some environmental pollution

Form of the lead	Principal uses
The element	Piping of water supplies, roofing
Alloys	Solders, bearing metals, type metal
White lead (basic lead carbonate)	Pigment in paints
Chrome yellow (lead chromate)	Pigment in paints
Red lead (oxide of lead)	Pigment and anti-rust agent
Lead silicate	Glazes
Lead tetra-ethyl	Additive to petrol

NOTES

1. Lead tetra-ethyl is a fairly volatile liquid at ordinary temperatures; all the other forms are solids.
2. Most of the forms of lead listed in the table are essentially insoluble in water; soft waters containing dissolved air and carbon dioxide can however dissolve significant quantities of lead; waters containing nitrates and organic acids of a peaty nature can also dissolve appreciable quantities of lead. The solubility of lead in water is much reduced in hard water due to the presence of calcium and magnesium salts; the pigments also tend to be slightly soluble in water containing dissolved carbon dioxide.

and soils is moved by the action of weathering and erosion, finding its way into rivers and eventually into the oceans.

The modern applications of the element and its principal commercial compounds are listed in Table 6.2. The element is still an important component of various commercial alloys, such as type metal (containing up to 94% of lead with added tin and antimony), bearing alloys (with up to 8% lead plus tin, antimony and copper) and soft solders (with up to 38% lead plus tin). Of these various alloys only solder which may be used in the sealing of tin cans might cause some lead contamination of food. The lead in solder is slightly soluble in some food acids, and some canned foods might have slightly higher than average lead contents. A similar problem may occasionally arise from a compound of lead used in lead glazes for earthenware vessels; these glazes containing lead have relatively low melting points and therefore require a lower firing temperature making them more attractive to handicraft potters.

Although many of the processes and the end products of the lead industry may cause additional lead intake by small groups of the population, any general exposure of the whole public can arise from four major activities.

1. Industrial processing and especially smelting operations.
2. The combustion of fossil fuels, especially coal.
3. The use of lead arsenate as an insecticide spray especially in orchards.
4. The use of lead additives in petrol to improve the octane rating.

The major industrial operations are primarily a risk to the plant operators, to a lesser degree, members of the general public residing within the vicinity of the plant, and only to a much smaller degree the population at large. Operations on one modern smelting plant were suspended recently when blood levels of between 0·8 and 1·2 ppm were found in some of the workers, these levels being approximately two or three times the normal lead levels in blood. In another industrial plant, contamination on the clothing of workers or in their hair was transferred to children at home, with the result that the children had to be sent away from home for a time until their blood levels returned to normal. Individuals in the neighbourhood of a processing plant and especially smelters are likely to be affected by small particles of lead carried away from the plant in gaseous discharges. A further source of environmental contamination may also arise from discharges of liquid waste into drains and the entry of lead into surface waters, estuaries and coastal waters.[2] The extent and the magnitude of lead contamination from industrial operations is difficult to assess in view of the lack of relevant data. This also applies to the amounts of lead released by the combustion of coal, but as reliance on coal is declining and also as improvements such as better filtration in industrial process plant are introduced, any local and general contamination from these sources should tend to decrease and contribute only a small part of the general lead pollution.

A rather surprising result reported by the Lead Panel of the US National Research Council Committee on the Biological Effects of Atmospheric Pollutants is the high lead content found in the dust and grit in the sweepings from the streets of American cities.[3] A value as high as 2650 ppm lead in the dust has been reported and a child would only need to consume about 40 mg of the dust daily to raise the blood lead level in excess of the normal value between 20–30 μg/100 g. The source of the contamination was not positively identified but may have been due to a combination of factors arising from fuel combustion and industrial activities.

General environmental pollution due to lead arises mainly from the combustion of petrol containing lead additives and, to a lesser degree, from the use of lead arsenate in orchard sprays. The use of lead arsenate is declining with the advent of the newer organic pesticides, but is still employed in several countries for the control of insect pests such as the codling moth, especially on hard fruits, apples and pears; this treatment is applied when the fruit is for export to countries where freedom from specified pests is required. The use of high octane rating petrol containing organic lead has risen steadily in the post-war period and is aggravated by the steadily rising number of cars on the road. An example of the increasing numbers of cars on British roads is presented in Table 6.3, an increase which is almost certainly matched by comparable figures elsewhere.

Table 6.3
Vehicles on British roads[4]

Year	Total no. of vehicles (millions)	No. of cars (millions)
1947	3·5	2
1957	7·5	4
1967	14·0	10
1970	16·0	11
2000 (est.)	34·0	29·0

Lead, in the form of its tetra-ethyl compound, was first added to petrol about 1925 to improve the octane rating and to act as an anti-knock agent. During the period covered by the table the use of high octane petrols has increased substantially, rising from about 3150 ktonnes in 1955 to 11,000 ktonnes in 1970; this rise has been accompanied by an increase in the lead content of the petrol averaging about

0·35 g/litre (1·7 g lead per gallon) in 1955 and rising to an average of about 0·54 g/litre in 1970. The highest lead content is found in premium grade petrol containing about 0·64 g/l in Britain in 1970. The total amount of lead added to all British petrols in 1970 would have been somewhat less than 10,000 tonnes, with a discharge to atmosphere from car exhausts of between 8000 and 9000 tonnes in the year. The total use of high grade petrol is considerably greater in the USA, where it has been estimated that the average American citizen uses about 21,000 gallons of petrol in his lifetime, and polluting the environment to the extent of about 50 kg of lead. The total petrol consumption in the USA is probably about twenty times that in Britain, and accounts for an appreciable fraction of the total annual discharge of lead to the environment from all sources, about 400,000 tonnes a year.

The lead in petrol is principally in the form of its organic compound tetra-ethyl lead, but after combustion it is largely transformed into inorganic compounds consisting of a mixture of oxides, chlorides and bromides, which are slowly converted into carbonate and basic carbonate as they diffuse through the atmosphere. Probably much less than 10% of the original lead is emitted in the organic form. As a consequence of the chemical transformation of the lead, it is discharged in the form of very minute solid particles of which about three-quarters are less than 2 μm and two-thirds are less than 0·3 μm in size. The smaller particles tend to remain suspended in air for some time and may be carried over considerable distances, whereas the larger particles will settle out more rapidly and hence closer to the point of origin. As a consequence of the settling pattern and the effects of air movements a concentration gradient is established with the concentrations of lead decreasing with distance from a busy arterial road. The amount of deposition will also fall off rapidly with increasing distance from the road, so that the heaviest contaminations of soils and of plants will occur close to the road. This effect is illustrated by some measurements of lead concentration in samples of lettuce grown close to Californian highways in 1965 (Table 6.4).

Table 6.4
Lead concentration in lettuce[5]

Distance from road	Concentration in lettuce, ppm	
	Unwashed	Washed
Less than 1000 yards	0·91	0·48
Greater than 1000 yards	0·51	0·21

The reduction in the concentrations with increase in distance and also after washing is consistent with a pattern of aerial deposition of

lead in particulate form. The general and widespread increase in lead
atmospheric levels since the 1940s is believed to be the basic cause of
the recent quite substantial increases in the lead concentrations of
Arctic snows and in elm tree rings (Fig. 6.1).

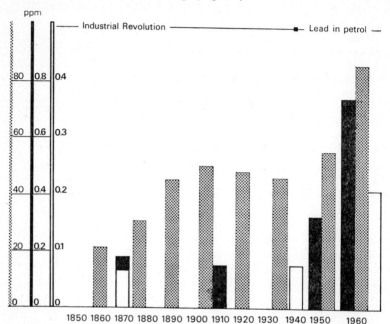

🔲 Lead on Scandinavian mosses ◼ Lead in elm tree wood ☐ Lead in Arctic snows

Fig. 6.1 Lead in the environment.[6] The figure illustrates the initial increase
in lead concentrations in the three types of sample starting about 1860 to the
turn of the century which is possibly related to increased mining of lead and
burning of coal. There is then a further sharp rise in the concentrations in all
three cases after 1950 which could be attributed to the increased utilisation of
leaded petrol. The data on the Scandinavian mosses have been obtained from
the paper by Ruhling, A. and Tyler, G., 'An Ecological Approach to the Lead
Problem', *Bot. Notiser*, 1968, **121**, 321–42, and the data on arctic snows from
the article by Murozumi, M., Chow, T. J. and Patterson, C. C., 'Chemical
Concentrations of Pollutant Lead Aerosols, Terrestrial Dusts and Sea Salts in
Greenland and Antarctic Snow Strata', *Geochim et Cosmochim, Acta*, 1969, **33**,
1247–94. The data on the elm tree rings has been obtained from the book by
Anthony Tucker, *The Toxic Metals*, Earth Island Ltd., 1972. For further
information on this topic with special reference to peat samples, the article by
Lee, J. A., and Tallis, J. H., Regional and Historical aspects of lead pollution
in Britain, *Nature* 1973, **245** 216–218, should be consulted.

LEAD IN FOOD

THE VARIOUS applications of lead create a twofold threat to human
health, firstly from the air which is inhaled and secondly from the food

and drinking water consumed. There have been numerous measurements of lead in the air of urban areas suggesting an average concentration around $2 \cdot 0$ $\mu g/m^3$. These average concentrations may increase almost tenfold near busy main roads but may only be about one-fifth in rural areas. The total intake of lead by the inhalation of about 20 m^3 of air by the average urban dweller may therefore be about 40 μg per day. The importance of this source of lead is difficult to determine as only about 10% of the intake may be retained in the lung tissues leading to a much smaller absorption. It is generally agreed, therefore, that food and drinking water constitute the major source of lead intake and despite a small absorption of the element through the g.i. tract.

It has to be admitted, however, that detailed knowledge of the transport of lead through the various natural food chains is strictly limited. Very little is known of the mechanisms of transfer of lead in either an aquatic or soil environment or of the concentration factors operating in the important food chains. The data which have been summarised in Table 6.1 suggest that lead is, however, concentrated in the higher species of a sea-water food chain.

e.g. CF for algae (algae/sea-water) = 28,000
      CF for fish (fish/sea-water) = 1,700

Lead concentrations in sea-water are normally extremely low owing to the insolubility of its compounds, rapid settling and hence a short residence time in the water but the figures suggest that some problems might arise in estuaries and coastal waters close to the sites of industrial waste discharges and sewage disposal.[2] In the soil environment the lead appears to be removed quite rapidly from the soil solution and is only available to plants in exceptionally small amounts from the majority of soils. There is also recent evidence to show that any lead absorbed by the root system is only translocated to a small extent to other parts of a plant with maximum concentrations remaining in the roots. The form of direct contamination of plants might have considerable significance in areas where atmospheric contamination is high as is illustrated

Table 6.5

Pb concentrations in cabbages grown close to a smelter plant in the UK[7]

| | Concentrations ppm fresh weight | | | | | |
| | Outer leaves | | | Inner leaves | | |
Sample	Untreated	Washed	Cooked	Untreated	Washed	Cooked
A	6·8	3·9	3·9	0·8	0·5	0·7
B	10·7	8·5	7·5	0·5	0·5	0·8

by typical results reported for cabbages grown in the vicinity of a smelter plant in the UK (Table 6.5). The concentrations of lead in the soil within a distance of 1·5 m east of the plant exceeded 100 ppm which is ten times the average natural level, but no correlation between the soil concentration and the concentration in the samples of cabbage could be detected. It will be noted that a much higher concentration is found in the outer leaves and that the lead content is again reduced by washing, both observations suggesting aerial deposition as the main source of contamination. One curious feature of the results is the apparent increase in the lead content of inner leaves after cooking, which the authors of the report attribute to losses of material having a low lead content during the cooking.

The existence of wide variations in the amounts of lead contamination in various items of food and the small number of dietary surveys for lead which have been carried out to date, make it difficult to assess average national dietary intakes of the element with any certainty. A recent and fairly comprehensive survey of lead in foodstuffs in the UK was undertaken by the Association of Public Analysts jointly with the Local Authorities Organisations.[8] A selection of the data obtained in this survey is presented in Table 6.6. The largest numbers of samples are for apples and pears, in view of the possibility of lead arsenate residues on the fruit imported from abroad. The results of a special survey of apples imported into the UK from 1964/5 crops are summarised in

Table 6.6

Lead residues in food, UK, 1967–8[8]

Group	Food sample	No. of samples	Pb concentration ppm Maximum	Estimated mean
Cereals	Bread	5	0·35	0·17
Meat, fish	Pig meats	9	0·50	0·16
	Poultry, game	1	0·30	0·30
	Sausages	6	0·35	0·19
Eggs	Eggs	2	<0·10	<0·10
Fats	Lard, dripping, butter	3	0·30	0·17
Fruits, preserves	Pears	71	0·20	0·10
	Apples	34	0·60	0·12
	Grapes	8	<0·10	<0·10
	Plums	4	0·25	0·14
Root vegetables	Potatoes	3	<0·10	<0·10
Green vegetables	Lettuce	6	<0·10	<0·10
Milk	Milk	3	<0·10	<0·10
	Cheese	9	0·60	0·16

NOTE

The mean value has been estimated by assigning a value of 0·1 ppm to all those results reported to be less than the limit of detection of the analytical procedure

Table 6.7

Lead residues in apples, 1964–5, crops imported into the UK[9]

Country of origin	No. of samples	ppm fresh weight		
		Core + peel	Flesh	Whole fruit
Australia	45	1·55	0·1	0·6
Canada	51	2·55	0·15	0·75
Italy*	5	0·35	0·3	0·35
South Africa*	5	0·15	0·1	0·1
New Zealand	6	2·1	0·1	0·75

* Lead arsenate not used as far as is known

Table 6.7. Comparison with results for earlier years shows a reduction in the lead residue levels of fruit mainly due to decreasing application of this particular insecticide. The total number of all food samples found to contain lead in excess of the limits of detection was small, and there was a considerable variation in the values reported. A similar wide variation also exists in results from the USA (Table 6.8). In these

Table 6.8

USA survey of lead content of some foods and beverages[5]

Food	Source	No. of samples	Lead content, ppm
Cabbage	Market	4	0·10 − 0·24
Wheat bread	Market	8	0·02 − 0·16
Bran flakes	Market	2	0·14 − 0·15
Spaghetti (prepared)	N.A.	2	0·06 − 0·21
Cornstarch	N.A.	4	0·75 − 1·83
Beef bone (fresh)	Western USA	1	3·60
Beef liver	Western USA	2	0·29 − 0·40
Beef kidney	Western USA	2	0·12 − 0·38
Beef cooked	Market	9	0·003 − 0·63
Eggs	Market	6	0·003 − 0·12
Lobster	USA	N.A.	2·5
Lobster	India	N.A.	0·08
Coffee (prepared)	N.A.	2	0·01 − 0·03
Beer	N.A.	3	0·13 − 0·29
Grape juice	N.A.	7	0·04 − 0·40
Wine	N.A.	10	0·05 − 1·51
Milk	USA – 1936	N.A.	0·02 − 0·04
Milk	Japan – 1936	N.A.	0·01 − 0·06
Milk	Cincinnati, recent	N.A.	0·04 ± 0·017
Milk	Northern and Eastern USA recent	N.A.	0·05 ± 0·025
Cocoa	Market (20 brands)	25	0·40 − 11·5

N.A. = Not available

circumstances, the estimation of a dietary average is very uncertain and can only be used to provide a very general guide to the contamination levels.

More recently still determinations of the lead content in samples of the seven major food groups as used in the dietary surveys for pesticide residues and mercury have been reported by the UK Working Party on the Monitoring of Foodstuffs for Heavy Metals.[10] The results obtained by the Working Party and estimates of the daily dietary intake of lead are presented in Table 6.9. It will be noted that the average concentration in the total diet is about 125 ppb. The average USA consumption

Table 6.9

Estimated human dietary consumption of lead[10]

Food group	Consumption % w/w	Average* Concentration of Pb, ppb	Consumption μg Pb/day
Cereals	16	170	44
Meat, fish	13·5	170	37
Fats	3	80	4
Fruits, preserves	13	120	24
Root vegetables	17	200	53
Green vegetables	7	240	24
Milk	29	30	14
Misc.	1·5	−	−
Totals	100 (1·55 kg)	125	200

* Average concentrations based on measurements in 14 towns from the fourth quarter 1970 to first quarter, 1972.[10] UK Regulations specify a limit of 2000 ppb for most foods

of lead has been generally estimated to be about 300 μg/day, rising to a possible maximum of 400 μg/day. The results for average consumption in the UK appear to be about two-thirds of these USA values. The major dietary sources appear to be fruits, animal products and cereals, with a decreasing contribution from the first item. The average results are perhaps indicative of the normal intakes of dietary lead but are expected to be higher for smaller groups of the population living close to industrial sites and urban highways. The above average values may also be expected for small groups of the population subsisting to a greater extent on shellfish with an average lead content estimated to be about 1000 ppb. Shellfish in particular appear to have the ability to concentrate lead from sea-water as in the case of mercury and other heavy metals, but the evidence suggests that it does not accumulate to a great extent in fish. The average lead content in canned baby foods at an average value of 240 ppb may be of some concern being almost double the concentration in the adult diet.

Any estimates of the average total intake of lead from all sources are further complicated by the uncertain contribution from the inhalation of contaminated air. As already mentioned, food is the major source of lead intake but there may be an additional intake from air of about 25% in rural areas increasing possibly to as much as 50% in urban areas with a high traffic density, and also drinking water.

As in the case of the other metallic pollutants in food, described in the previous chapter, there is again insufficient data from earlier years to enable any reliable estimate of the changing levels of lead consumption over recent years to be made and hence to distinguish any additional amounts deriving from changing patterns of lead use. Almost certainly there has been a small general increase in food contamination from the lead emitted by car engine exhausts, but the major increase in human lead levels will then be due to the inhalation of contaminated air.

Some recent measurements of lead in human hair have indicated that the total human exposure to lead despite the emissions from car engine exhausts may have declined since the beginning of this century.

LEAD IN MAN

THE LEAD which is ingested from food and drinking water is mainly in an inorganic form, and only about 10% of the daily intake is absorbed through the gastro-intestinal tract. The greater part of the lead intake from food is, therefore, eliminated directly and quite rapidly in the faeces. There is some evidence that a rather larger fraction of lead may be absorbed initially but is returned to the g.i. tract via the liver and bile duct. The remaining lead after passing through the liver is absorbed into the bloodstream to be transported and widely distributed throughout the tissues of the body. The transport of lead in blood, as in the case of absorbed mercury, is assisted by its ready association with protein and its preferential binding to the sulphydryl group of certain amino-acids in proteins. Further quantities of lead are partially eliminated in the urine, sweat and hair, but almost one-half of the amount absorbed is probably retained in the body up to a certain age. A large proportion of the retained lead is held in the mineral bone where it may partially replace calcium. The total amount of lead, therefore, is far from uniformly distributed in the body tissues with the highest concentrations being found in the bone (7–11 ppm), followed by the liver (ca. 2·0 ppm) and kidney (1 ppm) and with quite low concentrations (ca. 0·01 ppm) in the brain, heart and muscle tissues. The average concentration in the whole body ranges from about 1·1 to 1·5 ppm with a whole body burden between 80 and 120 mg. The amount of lead in the body continues to increase after birth, probably reaching maximum values in the brain and nerve tissues during the first two decades of life and in the bone after about four decades. Lead is also known to pass through the placenta so that a small quantity of lead is accumulated in the pre-natal period. The presence of higher concentrations of lead in baby foods may also be important in increasing lead levels in growing children at an early age.

The rather high concentrations which are found in bone and liver indicates a concentration factor from the diet to the tissues. Using the lead concentration of 125 ppb in the total diet, the concentration factor from diet to bone is about 70 and the corresponding concentration factor from diet to liver is about 15. These concentration factors are, however, exaggerated to some extent by the contributions from air and drinking water. On the other hand, there appears to be a discrimination factor of about 0·1 from diet to brain and nerve tissues. As in the case of strontium-90 it is of some interest to compare the lead to calcium concentration ratios in food and in the various body tissues.

e.g.   Lead units   LU in food $= 200 \ \mu$g Pb/1·1 g Ca
                                             $= 185 \ \mu$g Pb/g Ca
                      LU in bone $= 60 \ \ \mu$g Pb/g Ca
                      LU in the whole body $= 95 \ \ \mu$g Pb/g Ca

It is seen that the observed ratio from diet to bone i.e. LU in bone/LU in diet $= 0·3$ and the observed ratio from diet to whole body i.e. LU in whole body/LU in diet $= 0·5$. It should be noted that this discrimination factor against lead entering into bone is comparable with strontium-90 into bone (O.R. $= 0·25$).

The concentrations of lead in urine and blood are expected to bear some relationship to the dietary intake and especially to recent recurring exposures, although it is recognised that blood levels of lead are not a very sensitive indicator of changes in the intake. A total intake averaging about 300 μg per day is associated with average concentrations in urine of about 25 μg Pb/litre and blood concentrations of about 20 μg Pb/100 g in the average adult. Higher values than these have been encountered and in a recent survey of a group of Manchester schoolchildren some values were found to be more than double the average. Higher concentrations are also found in adults such as traffic police and garage workers and in all these cases it is clear that the additional lead derived from special exposures other than a dietary intake.

EFFECTS OF LEAD POISONING

LEAD, unlike mercury, in the environment is not converted into organic compounds and remains essentially in an inorganic form. The organic forms of lead are, however, far more toxic than the inorganic forms, but human exposure to the former will be limited normally to a very small contribution inhaled into the lungs from contaminated air. The inorganic forms of lead which are the source of human exposure from food share some of the properties of zinc, cadmium and mercury, described in the previous chapter. They bind quite readily with various proteins, probably by linkage with the sulphydryl bonds of certain amino-acids. This association with proteins assists the transport of lead to the various body tissues and is almost certainly a factor responsible for some of the toxic effects of excessive lead in the body. These toxic effects are generally expressed in three types of body cells; they are associated with irreversible changes in the brain and nerve cells, and also in the kidneys, and with reversible changes in the red cells of the bloodstream. Some interference with the normal enzyme processes of detoxication or potentiation of drugs and other foreign chemicals in the liver tissues may also be expected, although the magnitude and nature of any such effects is at present unknown. The lead which accumulates in the mineral bone is only slowly released and is relatively harmless until such times as it may be released during stress or in old age.

Nerve tissues appear to be especially sensitive to the adverse effects of lead accumulation; the brain appears to be protected normally (judging by the very low concentrations which are found naturally) as a result of the ability of the brain cells to discriminate effectively in favour of calcium. The precise mechanisms involved in brain damage are not known with certainty, but it is suspected that lead interferes with the normal biochemical processes involving those enzymes which release energy. It is also suspected that lead may interact more directly with calcium in the vital and complex sequence of events involving the transmission of nerve impulses. The effects on the brain and nervous tissues are especially severe and damaging in the case of young children. Anaemia, as a consequence of lead poisoning, is due partly to a reduction in the synthesis of the heme proteins (such as haemoglobin, myoglobin) in the developing red blood cells, the reticulocytes, and partly to a reduction in the normal lifespan of these cells (about 105 days) and a lowering of the red cell count. The mechanism involved in the shortening of the lifespan of the cell is not known. It is, however, well established that lead can interfere with the processes of heme synthesis at several stages each involving an enzyme reaction in the complex sequence of biochemical events. Interference with one important enzyme occurs at an early stage of the sequence of steps and involves the condensation of δ-amino-levulinic acid (δ-ALA) into a cyclical compound which is a vital stage in the formation of various porphyrins leading to haemoglobin synthesis. The consequence of this interference is the release of excess δ-ALA into the blood and urine, the analyses of which provide one of the earliest indications of exposure to lead well before the appearance of any clinical symptoms. There is evidence that the activity of this particular enzyme is reduced in proportion to the concentration of lead in the bloodstream and increases in the δ-ALA have been demonstrated at concentrations in excess of 40 μg lead/100 g of blood. The lead is also suspected of interfering with a much later stage in the synthetic route which, this time, involves the incorporation of the iron into the heme molecules.

The effects of lead on the kidney cells may again involve enzyme reactions which provide the energy for some of the transport and filtration mechanisms of this organ. These effects may impair the ability of the kidney tissues to reabsorb certain small organic molecules, such as glucose, and to excrete toxic substances such as uric acid. The latter may lead to accumulation of uric acid in the body with the typical symptoms of uric acid poisoning.

In any consideration of the long-term adverse effects of environmental lead upon the human population, any evidence of acute lead poisoning due either to industrial exposure or to sub-acute public exposure is not strictly relevant. The evidence for long-term chronic effects due to slightly increased lead concentration is, therefore, at present inconclusive and there are considerable difficulties in defining threshold levels below which it can be stated with confidence that no adverse effects will occur. As already mentioned, the earliest sub-clinical effects due to increased

Table 6.10

The harmful effects of lead in man[11, 12]

Lead concentration in the blood μg/100 ml	Estimate of regular intake to give blood level mg/day	Effect
20	0·3	Normal–no evident effect
> 40	1·0	Increase in δ-ALA in blood and urine
> 80	3·0	Reduction in blood red cell count; abdominal colic; long-term anaemia and progressive mental deterioration in children
> 120	10·0	Acute damage to brain and central nervous system

lead concentrations are changes in the concentration of δ-ALA in the blood and urine, which first become apparent at blood concentrations of lead in excess of about 40 μg/100 g. Some recent work suggests that this effect may be detectable at even lower blood levels. More serious and predictable effects due to raised levels of lead occur at blood concentrations around 100 μg/100 g (Table 6.10). At concentrations beyond 100 μg/100 g the effects of lead exposure become increasingly acute, involving symptoms such as abdominal colic with damage to brain and central nervous system, with symptoms such as convulsions, breathing difficulties, and loss of coordination, and damage to kidneys and liver. The blood concentrations of lead leading to chronic mental deterioration in children are much less than the values for adults.[13] There is no evidence for any long-term carcinogenic, mutagenic or teratogenic changes in man, although some cancers have been detected in the kidneys of mice and rats from chronic exposures to lead. Sterility, increased abortions and still births, and neo-natal deaths are also reported to be associated with acute lead poisoning in animals.

In view of all the uncertainties in determining safe levels for lead in the body, it has not been possible to establish, as yet, any ADI for lead in food and drinking water. If a normal concentration in blood of 20 μg/100 g is associated with a natural average daily intake of 300 μg/day, it is possible that the dietary intake could be increased to a level at which the blood concentration in the adult reached 30 or 40 μg/100 g without harmful effects. As the concentration of lead in blood is not related linearly to the dietary consumption, owing to the rapid increase in the excretion rate at higher intakes, a dietary intake of about 700 μg/day would probably be required to give a blood concentration of 30 μg/100 g. This is well above any normal variation to be expected at the present environmental levels of lead. A certain measure of control over lead intake is exercised by the application of tolerance levels to selected foods such as fruit and to drinking water. A maximum concentration of 3 ppm is allowed on fresh fruit and a maximum concentration of 0·05 mg/litre is allowed in drinking water in the UK.

REFERENCES

[1] *Lead Poisoning in Children*, J. S. Lin-fu, Publication No. 432, 1967, U.S. Department of Health, Education and Welfare. Social and Rehabilitation Services, U.S. Government Printing Office, Washington, D.C.

[2] *Pollution in Some British Estuaries and Coastal Waters*, the 3rd Report on Environmental Pollution by the Royal Commission, Cmnd 5054, HMSO, London, 1972.

[3] *Biological Effects of Atmospheric Pollutants*, Lead Panel of the US National Research Council, National Academy of Sciences.

[4] (a) *UK Vehicles on the Roads*, 1970, Department of the Environment, HMSO, London, 1971.
(b) *Annual Abstracts of Statistics No. 108, 1971*, Central Statistical Office, Government Statistical Service, HMSO, London, 1971.

[5] *Lead in the Environment and its Effects on Humans*, Bureau of Air and Sanitation and Environmental Hazards Evaluation, and by the Air and Industrial Hygiene Laboratory, California Department of Public Health, California, 1967, p. 39.

[6] (a) "Lead Pollution: a growing hazard to public health", D. Bryce-Smith, *Chemistry in Britain*, 1971, **7**, 54.
(b) "Lead in Arctic Snow", Jon Tinker, *New Scientist*, 1971 (2nd Dec.) 49.

[7] *Investigations into Lead Pollution of Surface Vegetables and Soils*, Reports of the Michaelis Nutritional Research Laboratories, Harpenden, Herts.

[8] *Joint Survey of Pesticide Residues in Foodstuffs Sold in England and Wales, 1st August, 1967 to 31st July, 1968*, published for the Joint Committee by the Association of Public Analysts, London, 1971.

[9] "Arsenic and Lead Residues in Imported Apples", W. D. Pocklington and J. O'G. Tatton, *J. of the Science of Food and Agriculture*, 1966, **17**, 570.

[10] *Survey of Lead in Food*, Working Party on the Monitoring of Foodstuffs for Heavy Metals, 2nd report, HMSO, London, 1972.

[11] "Lead", R. A. Goyer and J. J. Chisholm, in *Metallic Contaminants and Human Health*, ed. Douglas H. K. Lee, Academic Press, New York and London, 1972.

[12] "Diagnosis of Inorganic Lead Poisoning", a statement by various doctors, *British Medical Journal*, 1968, **4**, 501.

[13] "Behavioural Effects of Lead and Other Heavy Metal Pollutants", D. Bryce-Smith, *Chemistry in Britain*, 1972, **8**, 240.

Suggestion for further reading
Health Hazards of the Human Environment, WHO, Geneva 1972, chapters 3 and 12.

Addendum
The 16th Report of the Joint FAO/WHO Expert Committee on Food Additives (Technical Report Series No. 505, 1972), contains the following recommendation for "provisional tolerable weekly intakes" for lead:

3 mg per person, equivalent to 430 μg per day, and a food concentration of about 300 ppb.

7

Perspectives and Prospects for the Future

INTRODUCTION

THE EARLIER chapters have attempted to survey the magnitude of food and human pollution in terms of the principal groups of polluting agents. In each case, an attempt has been made to present quantitative estimates of the overall daily intake by the population-at-large from the principal dietary food groups. The average intakes have been supplemented where possible with information for small special groups within the population who are liable to be exposed to above average intakes. The distribution of the agent in the human body and the consequences of its presence have also been considered. In many cases, a reasonable estimate of average daily intakes is possible but the interpretation and significance of these values in terms of human health is an extremely complicated matter. It is very difficult to assess the adverse effects of chronic exposures with any degree of confidence and Table 7.1 presents a qualitative assessment of those agents which are involved in the more serious problems. The evidence linking the extent of food pollution directly with the amount of damage to man is extremely limited, and confined mainly to the several outbreaks of mercury and one of cadmicum poisoning. In nearly all other cases, the evidence for possible harmful human effects has to be deduced from experiments on animals. On the basis of these animal experiments, the table attempts to identify the more critical areas which call for further human investigation and evaluation.

The whole problem of the pollution of food and of man can really be reduced to three basic questions:

1. Is there a real food pollution problem? (Is it possible to diagnose any effects due to food pollution on the evidence available, and are there sufficiently reasonable grounds for assuming the probability for such effects?)
2. How serious is the problem? (Are the risks resulting from exposure to food pollutants in any way justified by the possible benefits of the application?)

Table 7.1

Food pollution: a qualitative assessment

The assessment categories:

1. Presence in the environment   A   Completely absent from natural background
   B   Additional to a background of similar substances
   C   Additional to a natural background

2. Presence in human food   A   Positive in all foods
   B   Positive in certain foods only
   C   Probable in all foods

3. The human risk   A   Children predominantly
   B   Special population groups
   C   The total population

4. Relative risk: ***   High
   **   Moderate
   *   Low

Qualitative assessment:

Source	Pollutant Group	Example	1 Presence in the environment	2 presence in food	3 Human risks
Agriculture	Pesticide chemicals	DDT	A	A	C ***
		2, 4-D; 2, 4, 5-T	A	C	C *
	Fertilisers	Nitrate	C	B	A **
	Antibiotic residues		B	B	C *
	Hormone residues	DES	B	B	C **
	Pathogenic micro-organisms	Salmonellae	C	C	C **
Industry	Heavy metals	Cd	C	B	B **
		Hg	C	B	A, B **
		As	C	B	C *
		Pb	C	C	A ***
	Radio-active fission products	Sr90	B	A	A ***
		Cs137	B	A	C ***
		PCBs	A	B	C ***

NOTE

The examples of the various types of pollutant represent only a small number of a much greater number of potential contaminants

3. Can any steps be taken to minimise or to eliminate any adverse effects?

The questions are clearly far easier to pose at the moment than they are to answer. There is insufficient evidence at present to define accurately the magnitude of any particular problem but there are good grounds to give rise to serious concern. The problem is far from being a simple one involving as it does the interaction of a multiplicity of factors, and probably quite beyond the capability of any one person to comprehend in its entirety. Probably the best that can be done at present is to try to identify the more urgent human problems (regarding which it is possible to report that some progress has already been made) and to, perhaps, stimulate a greater sense of urgency in arriving at an assessment of the overall situation. It is not possible to make any claims for a complete solution to the present problems and this chapter must therefore serve only as a contribution to their discussion.

Any analysis of the total problem involves the following considerations.

1. *The external environment* which is concerned with the expansion of industrial technology and the revolution in agriculture as the major sources of food pollution. The external environment also contributes to varying the spatial distributions of the effects of pollution which are in some cases on a global scale and in other cases on a local scale.

2. *The internal environment* of man which is concerned with any ultimate effects of the contamination in the food consumed and to which changes in dietary habits may also be important contributory factors.

3. *The time scale* which is concerned with the increasingly rapid rate with which changes are taking place in both the human external and internal environments, and the ability of the human organism to adapt to any possible effects. The temporal factor is, therefore, concerned with the increasingly rapid rate of changes in these environments, and an accelerated rate of change in almost a single generation.

4. *The future control measures* which are concerned with improving the methods of detecting and testing for the effects of food pollutants, any new approaches to reducing environmental pollution, and the possible provision of an efficient watch-dog service.

There can be no doubt that the deteriorating situation in the external environment has claimed a major share of attention to date and has captured the public imagination to the extent that public and political pressures of restraint are being exerted on the industrial producers of pollution. The consequences of the rapid evolution in agricultural methods may be more serious however in the long term, perhaps more difficult to reverse, and with the exception of pesticide residues have not captured the same degree of attention as industrial pollution of the environment. Any changes which are directed to improving the external environment can also be beneficial in the long term, leading to reductions in the contamination of food and the human internal environment.

The effects of any agent deriving from the external environment depend on its chemical nature, its route of entry into the body and its subsequent metabolism and distribution. The effects will also depend on interactions involving other foreign substances. A further complication in assessing the effects of dietary contamination is introduced by any additional sources of intake such as the inhalation of contaminated air. This is a special problem for example in the case of lead compounds.

Once the agent has been absorbed from the food and distributed within the body, the problem is then transferred entirely to the internal environment, which to date has not received anything like the same degree of attention as the external environment. Only a few voices have been raised so far in protest against the pollution of the internal environment and this may be the result of the attention which has been diverted largely to the problems of the external environment. There can be no doubt, however, that the problems of internal pollution have a complexity which is not matched in the external environment, they are more immediate and of intimate concern to the individual person and his descendants, and as such they call for serious attention and a careful objective appraisal. In this connection, it is important to appreciate that all foods contain a variety of organic and inorganic substances which are the body's essential nutrients. There are many other chemical substances present in very minute amounts whose precise role still remains to be identified. Many chemcal substances which are natural and known to be toxic may also be present in food, producing a variety of physiological responses in man, if they should be present in excessive amounts in the diet. These include substances such as 3, 4-benzpyrene which is present in many edible plants, cyanide in certain fruits, legumes and cassava, enzyme inhibitors such as trypsin inhibitor in soya beans, polyphenols (gossypol) in cotton seed oils and meal, thioglucosides which may act as goitrogens in brassica plants, ergotine in mouldy rye, and aflatoxins in mouldy ground nuts. The amounts of these substances are normally quite insufficient to produce any acute symptoms of poisoning, but any possible long-term chronic effects due to their presence in food at very low concentrations cannot be accurately assessed. To all these natural chemicals must now be added a host of other chemicals introduced deliberately as food additives. The problems of the internal environment are even further complicated by the increasing medication of the general public by a growing armoury of medical drugs.

The total number of substances in the form of chemical contaminants from the external environment, natural food poisons, food additives and medical drugs all entering into the human internal environment must be considerable and probably in the region of one thousand. Many of these are of very recent origin and although the body has tremendous powers of adaptation this ability must be related to the nutritional adequacy of the diet, the state of health of the individual and the rate at which these fresh exposures are taking place. The body's response to any chemical substance will, therefore, depend on a variety of host factors and these will also include variations due to the age, sex and race of the person. There may also be an inherited predisposition towards certain states of health as a result of certain inborn metabolic difficulties which may be aggravated by the presence of pollutants. It cannot, therefore, be too strongly emphasised that, in addition to the variations in the exposure of individuals through spatial non-uniform distribution of the diet and its contamination, individual human beings will also vary markedly in their responses to the effects of the various agents.

The external environment which is exemplified by an expanding technology and changing patterns of agriculture is the source of the modern range of chemical and physical agents of contamination. These substances are either present accidentally in the foods consumed by man or they may be added deliberately in food processing and preservation, or even utilised directly as medical drugs. The external environment imposes different patterns of food contamination depending on the extent of industrial development and its distribution within a country; it may also be responsible for regional and national differences in dietary habits. The latter are especially important in determining the state of the internal environment and the response of the host to any possible adverse effects from contaminating agents which are ingested. The total internal environment of the host organism is supremely important in this connection; its relevance is, however, being recognised increasingly as is evident, for instance, from the work and publications of one of its advocates, Dr R. J. Williams: "I predict that in the near future there will be a vastly increased interest in the internal environment, paralleling the surge of national and worldwide interest in external environments (e.g. pure air and water) which we have seen in recent years".[1] The importance of the internal environment will be considered in more detail in the third section of this chapter.

The external environment as the donor, and the internal environment as the recipient of any contamination, have also to be considered in relation to the important temporal factor. This concerns especially the ability of the organism to adapt to the consequences of the technological expansion when the effects are both numerous and occurring on a rapidly contracting time-scale. The changes in the internal environment may be latent and difficult to detect, but they could have a significance perhaps comparable with the effects of the changing external environment which have been identified with various stresses and strains in the human being. The historical role and the contracting time-scale of changes in the environment is therefore considered in some detail in the following sections.

The complicated responses of the organism and especially interactions within the organism suggest that it may be necessary to find alternative procedures for adequately testing the properties of environmental pollutants. This matter is also referred to in the concluding sections of the chapter, when various suggestions and developments which have been proposed to relieve the worst problems of pollution will also be considered in further detail.

HISTORICAL PERSPECTIVES: POLLUTION IN THE EXTERNAL ENVIRONMENT

THE HISTORICAL role of changes in human circumstances which affect both the sources of pollution and the foods consumed must be a key factor in any attempt to assess the significance of pollution for man today. These changes have normally been taking place slowly over a considerable time span ever since man first began to re-settle in Europe some 15,000 years ago after the last of the Ice Ages. For most of this long period of time, the population was extremely small and its activities would have created absolutely the minimum disturbance to the earth, whether through mining operations or cultivation. For some thousands of years even before recorded history, therefore, Europe would probably have seen the gradual settlement of a small peasantry "bent to the unchanging cycle of the seasons" and devoted to the simple tasks "of sowing, ploughing and reaping, tending the ox, the goat, the sheep and the pig, practising with such skill as they might command the arts and crafts of weaving and building, carving and pottery".[2] The early historical period would also have witnessed a gradual transition from simple weapons and equipment based on stone to iron, bronze and lead.

(a) The Industrial Revolution

THROUGHOUT THE whole of this early period and until comparatively recent times, the demands on the environment remained at an extremely low level. The first great change in human circumstances was quite a sudden affair, with the impact of the Industrial Revolution starting in the early 19th century. The Industrial Revolution was to change the population of England from an agrarian to an urban society; in the year 1801, out of a total population of 9M, just over one-quarter resided in the towns and three-quarters were widely dispersed in the countryside. A century later, the population had multiplied fourfold with three-quarters then living in the towns and cities and only one-quarter dispersed in the country. The Industrial Revolution was based on the rapid developments in science of the previous century and the growing confidence and ability of man to harness the results of basic science to modern technology. This acceleration in the rate of technological change, based on scientific discoveries, is illustrated well by the rapidly decreasing interval of time elapsing between an original discovery and its technological application. Light sensitive chemicals were discovered in the 17th century, but more than one hundred years were to elapse before photography became an established business. The principle of converting sound into electrical impulses was discovered in the 19th century and only about fifty years were required for its application to the development of the telephone service. In the 20th century, only an interval of five years was required to perfect the atomic bomb after the discovery of nuclear fission in 1939, and only a further ten to fifteen years were needed for the development of large scale nuclear reactors for generating electric power.

The second spurt in the Industrial Revolution followed on the end of the Second World War and witnessed the extremely rapid developments in nuclear technology, in the chemical industries, and in space technology. The emergence of the chemical industries was, indeed, a feature of the first Industrial Revolution. For the first time in recorded history, man then acquired gradually the ability to manipulate his chemical environment on an expanding scale. These large-scale developments of the chemical industries enabled the production of many natural chemicals for specifically human purposes, and these had previously only been present in the environment and interacted with man on a very modest scale. The production of artificial chemical fertilisers illustrates the rise to prominence of a small number of naturally occurring chemicals. The chemist is also able to create and to develop methods for producing a great variety of purely artificial chemicals having no natural occurrence and posing very different types of environmental problems. The manufacture of polymeric substances for the extensive range of modern plastics, many of the processes being based on an extension of the petroleum industries and the production of pesticide chemicals, such as DDT, are all excellent illustrations of this new capability. As a consequence a great many chemicals, either as intermediates or as end products of the chemical industry, were able to enter the environment and to create many of the modern pollution problems.

The great chemical expansion of the post-war years was also accompanied by even greater ingenuity in the production of an increasing range of synthetic chemicals, and an expansion in the fine chemical and pharmaceutical industries. A considerable number of these synthetics have been used as food additives at one time or another. It has, for example, been estimated that the average American is consuming about

5 lb of food additives every year, which is equivalent to a total consumption of about 500,000 tons per annum by the whole public. Several hundred of these substances are currently licensed for use in America and many of these have been introduced in the last 25 years. The actual number in use, at any one time, in significant quantities, will of course be considerably fewer. They are added to foods for a variety of reasons including preserving and extending the shelf-life, increasing their attractiveness by improving the appearance to the eye and taste, and maintaining a satisfactory texture. The chemicals which are added include the following groups of agents:

> Flavouring agents and condiments
> Colouring agents
> Preservatives, such as anti-microbials and oxidants
> Emulsifiers
> Sweeteners
> Vitamin and mineral supplements
> Flour improvers and modified starches.

Just as in the case of some of the environmental pollutants, some of the chemicals which are used are present quite naturally in foods, but others are entirely artificial and not normally present; many of them add nothing to the nutritional value of the foods.

The modern trend towards convenience and packaged foods is another factor which may also lead to contamination of food. Modern plastics are based on a number of long chain polymers, such as polyethylene and polyvinyl chloride (PVC) which are basically chemically inert and would not themselves contribute significantly to the contamination of food. The contaminating agents in this case are residues of chemicals employed in the polymerisation process or added subsequently to improve the appearance and the qualities of the plastic for the variety of uses to which the finished product is applied. These additives include chemicals employed as plasticizers, colouring agents, stabilisers to heat and light, fire retardants and anti-oxidants. It is these agents together with process residues, e.g. vinyl chloride monomer in the polymer which may migrate slowly into the foods from the packaging materials adding to the other chemicals which may already be present.

In addition to all these substances and the wastes of industry, other problems of human contamination are created by the pharmaceutical industries with their increasing ability to modify natural drugs to increase their specificity in medicine or to create entirely new synthetic drugs. Some indication of the magnitude of this problem is also illustrated by data from the United States. In 1966, for example, it is reported that 167M prescriptions were issued for a range of medications which included stimulants, sedatives, analgesics and tranquillisers. In a small survey, conducted in 86 households in San Francisco, there was an average of 30 medicaments per household of which about one-fifth were obtained on prescriptions. It has also been estimated that about 9M doses of amphetamines and about 40 tons of barbiturates were also being produced annually in 1962. The problem with these particular chemical drugs and others, such as LSD, is that probably about half of these quantities are entering into the illicit channels of distribution. (See, for example, *Drugs of Abuse*. ed. S. S. Epstein, M.I.T. Press, Cambridge, Mass., 1971.)

(b) The Agricultural Revolution

THE ESSENTIAL features of the agricultural revolution have been referred to in Chapters 2 and 3, which are concerned with the problems of the residues arising from large-scale applications of chemicals in farming. Modern developments in agriculture, which are still continuing, have taken place in a very brief period of time, which is probably no more than about 1 % of the total time in which agriculture has been practised. The main changes have involved the rapid intensification of methods of crop cultivation and animal stock rearing. The rapid changeover to homogeneous systems of mono-cultivation of crops specially selected for their high yield quality in a small number of larger-sized fields, coupled with the generous application of chemicals, poses the more immediate threat to both the external and internal environments. This situation is viewed with some concern by a growing number of scientists and has been well expressed by Dr Walter Ryder, formerly Head of the Entomology Department in the Institute of Animal Sciences of Havana: "Over the past few thousand years, man has developed a capacity now enlarging with unprecedented speed, to disrupt the processes of gradual biological evolution and ecosystem modification . . . It would seem logical to deal with the inevitable consequences (among them the problems of pests and diseases) on the principle of restoring to the environment some measure of stabilising heterogeneity."[3]

The green revolution involving the introduction of high yield varieties of cereal grains in a number of Asian countries has already made a notable contribution towards relieving those countries from near famine conditions and from their former dependence on imported grains. The latest annual review of the United Nations FAO, however, seems to imply that the initial rapid increase in crop production with the introduction of these special varieties is now slowing down. The 1971 rate of increase in production has been estimated at only about 1 % against an increase of about 4 % in previous years; the food output per person in these countries has actually declined owing to the more rapid increase in the population growth.[4] There are also fears that the new high yield varieties may have a lowered resistance to pest infestation and may require the continuing application of chemicals on a large scale for some time, or until such times as resistant strains are available and established.

The agricultural revolution has, therefore, been a major factor in introducing a wide range of pesticide chemicals into the human environment and the great majority of them in a very short interval of time. To these must be added residues of antibiotics and hormones used in intensive stock rearing.

HISTORICAL PERSPECTIVES IN INTERNAL CONTAMINATION

DURING THE whole of the period of man's gradual change-over from a nomadic to a settled existence dietary habits would also have been evolving gradually to satisfy essential human needs without any systematic knowledge of nutritional requirements. The ability of man to evolve a satisfactory diet even without the advantages of nutritional science has been demonstrated by the discovery in quite recent times of quite primitive and remote communities who have evolved for themselves a perfectly adequate diet based on a selection of locally available foods. To arrive at such a balanced diet in these circumstances probably involves the use of a long period of trial and error in which those foods which are beneficial are selected preferentially and those which are harmful are eliminated. Dietary habits which have evolved slowly to meet the instinctive needs of the people have only begun to change rapidly in very recent times. Harmful factors in the natural diets may not be entirely absent, but there would be an adequate time for the organism to adapt to any adverse factors.

The recent industrial changes referred to in the previous section have all been accompanied by the rapid urbanisation of the population. An immediate consequence is that more and more people are separated from the resources of the land and less dependent on their own methods of food cultivation and preparation. These changes are illustrated by the rapid expansion of bakeries at the expense of home baking at the early stages of the Industrial Revolution and also by the rapid development in the scale of mechanisation of food industries, such as biscuits, jams and confectionery. The development of bakeries was also accompanied by a switch from barley, rye and oats as the basic grain ingredients of the bread to wheat and a growing preference for white over brown bread. A similar trend was also observed in the growth of the breweries and the almost complete disappearance of home brewing. The coming of the railways helped to expedite the entry of fresh farm produce into towns and cities, but increasing reliance had to be placed on imports as the population expanded. There was, for instance, a considerable expansion in the amount of sugar consumed after about 1850. The most rapid changes in the average dietary habits have, however, been occurring since 1945 and accompanying the rapid changes taking place in the external environment. The most recent trends have involved the supply of an increasing number of refined and processed foods in cans or in packages, all designed to ease the tasks of food preparation in the kitchen. However desirable these changes may be, they place the consumer at one further remove from the original sources of food. The refining and processing of these foods demands a greater uniformity in the raw food materials, sometimes at the expense of quality, and frequently requires the introduction of the chemical additives to improve flavouring, palatability, texture, and shelf-life. Gross adulteration of food was not uncommon in the last century and the Food and Drugs Act of 1860 was introduced to exercise legal controls over the quality of the food offered to the public. The gross abuses of the 19th century are unlikely to be repeated; the problems have just become more complex and difficult to evaluate because of the large number of food additives which might be used at low concentrations.

The advantages of processing are essentially that necessary foods can be made available all the year round and that their nutritional quality and purity can, at least, be subject to measures of control. The disadvantages which have to be considered are that changes in the nature of the foods as a result of processing and the extent of the use of chemical additives may yet have long-term adverse effects on human health. The essential question relates to the ability of the human organism and especially the digestive system, which is the product of a long evolutionary development, to adapt entirely to the extremely rapid changes to which it is now being exposed.

The technological revolution in food processing and marketing in recent years and which is still gathering momentum, has taken place simultaneously with the other developments in agriculture, industry and medicine. The many chemicals which are present in food and in the body from these sources are additional to very many other substances which are present quite naturally. All natural foods are in fact quite complex mixtures of chemical substances including the major nutrients such as the carbohydrates, fats, proteins, mineral elements and other essential substances such as the vitamins and trace elements, natural colouring, flavour and aroma compounds present in small amounts. The chemicals which are used as food additives or are present as impurities, cannot be assumed to be automatically toxic to man and the essential problem is one of identifying those constituents, whether natural or artificial, which present long-term risks to the future health of the human population.

The many changes in food habits and the existence of food contamination are largely accomplished facts and may be very difficult to put into reverse. The human organism is extremely adaptable in coping with changes but it could certainly be argued that it is now being subjected to too many basic changes in too short a time. There are already sufficient warning signs that all may not be so well with the state of human health as we are sometimes led to believe. It is indisputable that developments in medicine and in public hygiene have extended the lifespan of the average individual, and have been particularly effective in reducing the incidence of infantile mortality. The annual death rates from infectious diseases, such as influenza, diphtheria, and tuberculosis, the major killers until recent times, have shown a progressive decline due to medical advances. Tuberculosis, for example, declined in the USA from about 2000/million of the population in 1900 to about 200/million in about 50 years. Despite this achievement, however, the death rates due to chronic causes, such as cardiovascular disease and cancers have shown a steady increase. In the USA, for example, over the same period of 50 years, the death rate due to all forms of cancer almost doubled from about 700/million of the population in 1900 and cardiovascular disease rose from 3400 to 5200/million population. In the case of cancers, there has also been a marked increase in the incidence of lung and

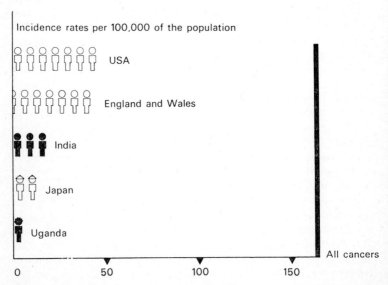

Fig. 7.1 Civilisation disease: cancer of the colon

colonary cancers which are now responsible for the majority of cancer deaths. The incidence of lung cancers is almost certainly due to non-dietary factors but the increase in colonary cancer may well be associated with changes in food habits. Recent epidemiological surveys into the incidence of colonary cancers in a number of countries including several of the developed countries with western standards of life and a number of the developing countries of the world suggest strongly that dietary habits are involved. The work of Dr Burkitt of the Medical Research Council has special relevance in this connection.[5] A selection of his results embracing the highest and the lowest rates of colonary cancer incidence are presented in Fig. 7.1. The high rates in the Western nations, which approach 500/million population per year, should be compared with the low rates in the developing nations, typified by Uganda, at about 50/million of the population per year. This marked difference may be associated with a preponderance of refined and pro-cessed foods in the diet of the Western nations compared with almost unrefined raw foods in the more primitive village communities of the developing nations. The role of the dietary factors was examined further by Burkitt by means of experiments involving small groups of children in an English boarding school, an African boarding school, and a Ugandan village. In each of these three small groups, the transit times for the passage of food through the alimentary tract were measured by incorporating small, totally inert, plastic pellets into the food. The transit times were also compared with the weights of faeces eliminated daily; all these results are presented in Fig. 7.2. They show striking differences between the groups, with the English boarding school-children having an average transit time of 90 hours and a small daily weight of faeces of just over 100 g, and the primitive community having a transit time of about 35 hours and an average weight of faeces of about 460 g per day. The African boarding school represents an inter-mediate stage of dietary development. These differences indicate strongly that dietary factors are having a marked influence on the digestive processes and one of the factors which might be involved is the amount of indigestible fibre or roughage in the diet. When these results are related to the epidemiological evidence of differences in the incidence of colonary cancers, they imply quite strongly that dietary factors have a dominant causative role. The precise factors and mechan-isms which are the direct cause of the cancer cannot yet be identified with certainty. The influence of roughage in speeding up the passage of food through the g.i. tract will certainly reduce the residence time of the non-absorbed portions of the diet in the lower digestive tract and in the large intestine, limiting the time available for any interactions between contaminants and the surfaces of sensitive tissues. The reduced contact time could, therefore, have considerable advantages to the

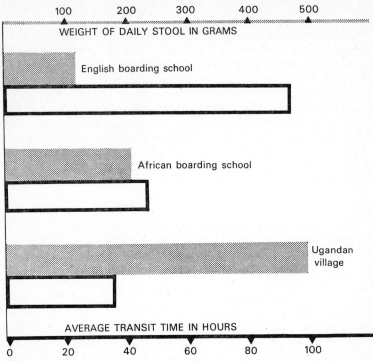

Fig. 7.2 Digestion of food: the transit times for the passage of food through the alimentary tract in three different groups

organism. As many agents are eliminated in the faeces, either by failure to be absorbed during their passage through the small intestine or after re-entry via the entero-hepatic cycle after being absorbed, this possibility cannot be discounted. A further possibility affecting the whole body is that the more rapid transit times will also result in a smaller uptake of any contaminating agent through the small intestine. Birkitt himself, however, considers that changes in the populations and strains of the colonies of micro-organisms inhabiting the large intestine, and the lower part of the small intestine may be more directly responsible. Although this cannot be confirmed at the present moment, it could have special significance when residues of antibiotic are present in food or when they are used in medicine.

The importance of Burkitt's work is the clear demonstration that dietary habits have a marked influence on the digestive processes in man; this may also subsequently influence the elimination and distribution of the contaminating agents within the body. Other dietary factors, in addition to the content of indigestible fibres, may also contribute to the problems of chemical contaminants and the health of the individual. One of these factors might be the considerable increase in the quantities of refined sugars and starches consumed in the average Western diet. The average sugar consumption in England has increased almost twentyfold in the last hundred years and Yudkin has presented evidence relating this increase to the growing incidence of cardiovascular disease.[6] Cleave, Campbell and Painter[7] have also reviewed the evidence for refined sugars and starches as contributory factors to cardiovascular diseases and the other civilisation diseases in Western countries. There seems little doubt, therefore, that one or more of these factors associated with changes in food habits and the composition of the foods consumed may well be associated with the rising incidence of chronic illness in the West. They lend weight to the central thesis of Dr R. J. Williams "that the nutritional micro-environment of our body cells is crucially important to our health and that deficiencies in this environment constitute a major cause of disease".[1] Williams is, of course, concerned primarily with the deficiencies in the nutritional quality of the diet but if these deficiencies are extended to include the various food contaminants, they may well reinforce his basic premise.

TESTING FOR TOXICITY

THE CONSEQUENCES of tissue exposure to the variety of contaminants or to their metabolic products at low concentrations after passage through the alimentary tract remain the central problems. Any estimates of the results of such exposures are almost exclusively based on studies involving a single agent and, more occasionally, involving pairs of small groups of agents. The great majority of the work which has been reviewed in the earlier chapters appears to reach at least a measure of agreement that the available evidence for enabling sound judgements in such a complex situation is either quite inadequate or just not available. "At the moment, we know practically nothing of the long-term effects on the body of an ever-increasing number of food additives and non-essential medications, and yet we are all exposed to this hazard which could change our physiology and change our disease patterns if allowed to continue unrecognised . . . Our ignorance is abysmal; even more frightening is our lack of regard for the problem." Professor Beaconsfield's concern is only reinforced by the presence of all the environmental pollutants.[8]

The problems of testing, with special reference to drugs and food additives, have been discussed in some detail in the paper by Professor Beaconsfield. It is, of course, well known that drugs after tests on animals are usually subject to clinical trials on closely observed human beings, but it is also apparent that it is impossible to design trials which are completely foolproof. The prediction of human toxicity from animal experiments is also subject to very serious limitations and especially when different reactions and metabolic patterns between species are to be expected. Professor Beaconsfield suggests that it is quite possible "to obtain a so-called normal physiological picture which would not reflect underlying physico-biochemical alterations", which could lead ultimately to irreversible damage, or side-effects in the exposed organs. These are a manifestation of disturbed biochemical and biophysical processes and it is imperative, therefore, that there should be a much closer understanding of the precise mechanisms of damage and that these should be studied at the tissue and cellular levels. "Such testing should tell us whether the cells of the body's vital organs are deranged functionally or architecturally by exposure to the compounds; whether any changes which do occur are cumulative and reversible on withdrawal of the compounds; and whether any cytogenic effects are demonstrable."[8] It is recognised that such tests present a formidable undertaking and that there is an urgent need for scientists of multidisciplinary training to undertake the work. Wherever possible, experimental evidence of toxic effects should also be backed up by epidemiological studies on small groups of a population selected for their known exposure to higher levels of the contaminant. The special studies which the World Health Organisation are undertaking at a few places which have higher than normal natural radiation backgrounds, such as at Kerala in India and at La Paz in Bolivia, are a good example of this type of investigation.

The limitations in predicting human toxicity from animal experiments are due mainly to variations in the response of the different species and their varying sensitivities to different agents. These variations may be due to differences in the metabolic pathways and to differences in tissue sensitivity to the agent or its metabolites. There is also the experimental problem of having to use large numbers of animals in order to obtain statistically relevant data. As an example, it may be necessary to use as many as 30,000 animals to demonstrate the incidence of an effect such as cancer which might occur in man at a rate of 1 in 10,000 from exposure to contaminated food. The only practical alternative to using large numbers of animals is to increase the level of exposure well above the expected levels in food. This procedure then involves a dubious double extrapolation from high to low exposure in the animal experiments and the further extension of the results to man himself. Studies of the genetic effects from exposure to ionising radiation provide an example of one of the very few large-scale experiments involving large numbers of animals. Even in this case there was still an appreciable gap between the lowest experimental exposures and the exposures due to food contamination.

What is therefore required at the moment is a comprehensive examination of the effects of the many agents already present in food, not only to extend the studies in progress on individual agents, but to develop the work to include groups of agents, and especially those groups where there are grounds to suspect that interactions may be occurring. The formidable nature of such a programme involving the present techniques of experiments on animals and their doubtful significance in terms of human effects requires a fresh approach to the whole problem. Although the alternative methods involving tissue cultures may also have their limitations when the effects have to be extrapolated to the whole organism they should be capable of revealing the mechanisms of damage and the nature of any interactions which are taking place at the cellular level. It may then become possible to make reasonably confident predictions of the consequences of human exposure and especially to screen the consequences of any new agent entering the environment. The task is however so formidable that to stand any chance of success it must command the support of governments and international agencies.

Any agent of pollution in addition to satisfying the first criterion of safety, assessed on the best available evidence, should also satisfy two other criteria of efficacy and identity.[9] The criterion of efficacy requires that any benefits in using a chemical agent must be sufficient to far outweigh any potential hazards. As there is an obvious element of risk in most human activities (*see* Chapter 1, p. 33), this striking of a balance between the benefits and the risks calls for careful and considered judgement, and should also have the acquiescence of the general public. This is obviously a difficult matter, but the difficulties should not be allowed to stand in the way of a properly informed public. The third criterion, that of identity calls for a clearly defined specification of the composition and purity of any chemical agent, perhaps on the same lines as the WHO International Pharmacopeia of drugs and medicines. This criterion is necessitated by virtue of the processes used in the manufacture of the chemical, and the probable presence of impurities which even in very small concentration may be more toxic than the principal agent. The presence of trace amounts of dioxin in 2, 4, 5-T illustrates this particular problem.

REMEDIAL MEASURES

ANY IMPROVEMENTS in the quality of the external environment should be very helpful in the course of time in reducing the risks of food contamination. These improvements will concern the manufacturing industries and especially those involved in the production of chemicals and the generation of power at the lowest practical cost. The products and wastes of these industries are accidental food contaminants and are referred to on p. 223. Modern agricultural practices are another source of food contamination and the problems of reducing this source are referred to below:

(a) Measures to control agricultural pollution

RESIDUES OF crop protection chemicals in food are a major source of present day human pollution. In addition to their threat to human health, there is also an increasing concern regarding the efficacy of their applications in agriculture and especially the insecticide chemicals. This concern arises for two different reasons:

1. The indiscriminate nature of the attacks of these chemicals on insects, both those which are pests and others which may be beneficial. The damage to honey bee colonies reported several years ago due to the persistent chlorinated hydrocarbons is one such example.
2. The emergence of certain insect pests with an acquired resistance to the effects of the chemical. When the insect pest is able to acquire this resistance sometimes more readily than other insects which may help to keep it under control, the pest may even become more firmly established than it was prior to treatment. There are now many examples of the emerging chemical resistance, one outstanding example being the malaria-carrying mosquito. For all these reasons, various alternative systems of protection such as more specific chemicals or biological controls are the subject of a great deal of present attention.

An effective long-term alternative to the present pesticide chemicals in the context of modern agricultural practice is not yet available and many of those which are under investigation may only be temporary palliatives. It is, therefore, reasonable and it is becoming increasingly practical to consider a genuine and perhaps very radical alternative approach which would not only relieve the worst aspects of crop pollution but would, at the same time, make a very positive contribution towards solving the intransigent world food problem which remains a matter of some urgency. Both of these problems could be effectively tackled by the direct conversion of agricultural crops, such as several varieties of bean and cereal into acceptable analogues of meat and milk, by-passing in the process the highly inefficient animal. The majority of the world's population already relies very heavily on plant sources for its essential protein as is illustrated in Table 7.2. The dependence of the world population directly on plant sources for body energy is even more pronounced. The P/A ratio for protein and the whole world

Table 7.2

The sources of protein in 1960[10]

Source		Protein consumption, Mtonnes/yr.		
		Developing countries	Developed countries	Whole world
Animal produce	(A)	6·6	14·8	21·4
Plant produce	(P)	34·7	12·6	47·3
P + A		41·3	27·4	68·7
P/A		6·6	0·85	2·2
Population		2·5G	1·00G	3·5G

It has also to be noted that the 5/7ths of the total population living in the developing nations has only about half of the total land available for cultivation and this is generally less fertile than the available land in the western nations. The average income of the population in the developing nations is only about 1/10th that of the western nations so that any attempt to follow western agricultural practices with emphasis on animal proteins can only be disastrous for the developing nations N.B. IG = 10^9 = 1000 million.

population is increased to 9·0 if a further 135 Mtonnes of protein for feeding livestock is included in the total. This is mainly because the average protein conversion efficiency of farm animals is extremely small and usually less than 10% (the average protein conversion efficiency is the percentage of edible protein produced by the animal over the protein which it receives in its feedstuffs). It is the huge demand for animal feedstuffs which is responsible for most of the enormous pressure on the land to increase its productivity. The various processes of converting crops into acceptable protein foods which are now readily available would operate at much higher efficiencies approaching 100%. This would relieve the pressure on the land very substantially and the necessity to force increasing production off it to satisfy the rising demand. The necessity to resort to intensive stock rearing would also be reduced and the risks from pathogenic organisms, antibiotic and hormone residues would also be drastically decreased. The analogues of meat and milk which can now be prepared from plant sources can be made perfectly acceptable and nutritionally adequate with the further advantage that the quality of the product can be very carefully controlled. A number of these products are already available on the market. The additional bonus is that food crops, originating sometimes in the developing countries, can be diverted immediately to their own urgent needs.

An immediate and further consequence of relieving the pressure on the land would be the opportunity to resume methods of crop cultivation based on more traditional systems of diversity and rotation. This is important in contributing to Ryder's principle of restoring to the environment some measure of "stabilising heterogeneity", and thereby to ensure that "in carving uniformity out of heterogeneous ecosystems" the point is not reached, "where the quality of life is bound to suffer"[3]. This can be achieved by the rotation of crops or by the deliberate interplanting of crops in the same field. A reduction of 50% in the incidence of insect pests in maize was attained recently in Cuba by sowing sunflowers in eight-metre strips in between the sowings of the maize. Heterogeneity in crop production may also encourage the introduction of some of the other alternative and less drastic methods of controlling the worst ravages of pests such as the biological control systems. These can assume a variety of forms, a classical example being the introduction of an Australian ladybird into the California orange groves to control the attacks of the citrus scale beetle which were quite serious at the beginning of the century. This proved highly successful until the fruit growers decided to introduce DDT about 1945; this proved to be more harmful to the predatory ladybird than to the beetle pest, which was enabled to re-establish itself in greater intensity than ever before. This example of insect control by a biological method is also matched by another example of plant weed control; this involved the introduction of an Argentinian moth into the Australian outbacks in 1930 to control the spread of the "prickly pear" and to preserve land for grazing purposes. Other systems of biological control which are being developed and in some cases tried out in the field, include viral and bacterial controls, and the release of male insects sterilised by an irradiation treatment in the pupal stage.

There may also be a need to continue with the application of certain chemicals subject to stringent regulations and safeguards. These could contribute to a state of heterogeneity in the insect population of the environment, and therefore play an effective part in an overall pattern of pest control. They could include some of the chemicals in use at the moment but possibly treated with protective coatings to prolong their lifetime in the field and ensuring that they become available only at the appropriate time; these could be supplemented by others of a more specific nature for the insect hosts which they attack. They could also include minute amounts of other special chemicals which are able to interfere with the weak chemical signals which direct the behavioural responses of the insects. These could be used to mislead an insect in the selection of the host plant for its food or for the deposition of its eggs; they can also be used to affect the sexual behaviour of the male insects.

The problems associated with the indiscriminate use of pesticide chemicals are now much better appreciated, and there has been an encouraging response from various governments in banning or in restricting the applications of some of the more persistent chemicals, especially aldrin, dieldrin and DDT. Another form of agricultural pollution, involving the contamination of animal produce by pathogenic strains of bacteria with acquired resistance factors to certain antibiotics used in human medicine, is also expected to show some improvement in the UK; this follows the government's decision to make it unlawful to sell feedstuffs containing certain antibiotics, such as penicillin and the tetracyclines, used in human medicine.

(b) Measures to control industrial pollution

INDUSTRIAL CONTROL of pollution is very largely a matter of economics and could be achieved in very large measure by increased expenditure on the treatment of the various forms of wastes using technical processes which are already available. The control of industrial discharges into the environment is therefore very largely a matter of prevention by adequate treatment at source. Some measures of control over industrial pollution and of preventing any substantial increases could undoubtedly be achieved by curbing the rate of economic growth and especially by restricting the rate of growth in the manufacturing and power industries. The manner in which the demand for power multiplies at a faster rate than that of the growth in population suggests that there will be considerable difficulties in realising any effective curbs on economic growth in the foreseeable future. The general public would clearly be reluctant to sacrifice any of the material benefits which it enjoys at the moment, and even if these were to be at the expense of reduced contamination of the environment. It has also been made perfectly clear that any attempt to restrict economic growth in the developing countries will be quite unacceptable to the countries concerned. It would appear therefore that the only effective controls will have to depend on technical measures for processing the wastes of industry and especially for eliminating any sources of toxicity in the wastes. This will increase the costs of production inevitably, and these will have to be borne by the public either through increased costs for the products of industry or through government grants with the costs shared by all members of the public. There is certainly no technological reason why the worse effects of industrial pollution such as waste effluents into rivers and coastal waters should not be brought rapidly under much greater control. The British government has recently announced vastly increased sums of money to clean up the rivers, and the necessary legal powers such as the Clean Rivers Acts of 1960 are also available. Another important measure is to restrict the amount of lead which is permitted to be added to petrol; the first steps towards this have also just been announced by the British government, although they may be expected to have only a marginal effect initially.

The growing dependence of the power industries on nuclear energy sources presents a number of special risks. These include the rather unlikely but potentially serious type of accident to a nuclear reactor involving the loss of coolant in the reactor circuit leading to a possible failure of the containment systems. This problem has given rise to much concern in the USA in recent years in the case of the Light Water Reactors. This type of accident involving a reactor and, to a lesser degree, an accident during the transport to the chemical separation plant of highly radioactive fuel elements discharged from the reactor cores could lead to locally high contamination levels from escaping fission products. There are also the problems of coping with the tremendous quantities of fission product wastes after the chemical separation process, a problem which is dealt with in the UK by concentrating them into a small volume and storing as liquid waste. The quantities of such waste will grow as the nuclear power programme develops and storage will be required for at least a hundred years because of the considerable amounts of long-lived fission products such as strontium-90 and caesium-137 in the wastes. Alternative methods of storage are being investigated such as conversion into solid glassy materials in which the active materials would be effectively trapped. These problems will remain so long as the nuclear power programme is dependent on nuclear fission, but will be very largely eliminated if and when nuclear fusion is able to take over, perhaps towards the end of the century.

There is also mounting concern for the activities of small terrorist groups and the possible hijacking of radioactive materials and especially the fissile element plutonium produced in ever larger quantities as the nuclear power programme develops. Even if this plutonium could not be diverted to the production of atomic weapons, it might be used to pose a threat to water supplies. It is evident, therefore, that strict security measures for the storage and transport of this material are absolutely essential.

(c) Further requirements for effective pollution control

THE EXISTENCE of legislation is in itself insufficient to control excessive pollution unless it is supported by powers of inspection and is backed up by the resolve of an informed public. Many of the legal powers to control the worst excesses of pollution in the UK have been available for some time; they may have been responsible for avoiding even worse pollution of rivers and estuaries but they did not succeed in preventing many of the abuses which exist today.

The effectiveness of any measures which are taken to control pollution at its source will still require enforcement through adequate systems of monitoring and inspection. The UN at the recent Stockholm Conference on the Environment agreed to the setting up of global systems of monitoring for certain pollutants, and these are almost certain to be backed up progressively by national arrangements. The regular monitoring of foods for radioactive fall-out and certain pesticide residues is well established in Britain, even though there has been a considerable reduction in the number of food samples checked regularly for radioactive fall-out. A scheme for the monitoring of some of the heavy metals in fish and other foods which are important in the national dietary has also been put into operation. All such measures are to be welcomed and encourage the hope that increasing information can help to pinpoint the major sources of pollution and provide invaluable assistance in planning the necessary control measures.

Many other suggestions have been made from time to time to improve the effectiveness of any steps to control pollution, and many of these are reviewed in the paper by Epstein.[9] They include such measures as:

1. The creation of international data banks or a registry of all chemicals which may pose problems of pollution; the chemical specifications of such compounds could also be included, in addition to what is known about their efficacy, toxicology, results of monitoring, and epidemiology. It might also be possible to integrate any such arrangement with an international regulatory and early warning alert system.
2. The enforcement of existing national legislation or the enactment of new legislation to impose the required standards of control, to ensure that all issues relating to human safety and environment quality, and all data relevant to such discussions, are made public, and finally to ensure full measures for impartial and competent testing of the sources.
3. The establishment of multi-disciplinary training courses and research programmes into the whole range of environmental problems.

The success of all attempts to improve the human environment, externally and internally, will depend ultimately on the will of the population expressed through the media and national governments. There can be very little doubt about the growing anxiety of the people, which it would be unwise to ignore. There can be no doubt whatever that there is also a serious need for much more information, to enable the whole range of problems to be analysed and discussed in a cool and rational manner, and not in the fragmentary fashion which has largely prevailed up till now. In the words of John T. Edsall of the Biological Laboratories of Harvard University, "I am not one of those who is crying out 'Doom within the Decade', but as far as keeping the world a livable place is concerned, I regard present trends as extremely ominous. As one example I suspect a survey of the world would show a steady increase over the last half century and more, in the amount of land that is becoming desert or semi-desert . . . I suspect that modern technology, and the need of feeding a rapidly rising population, are accelerating the process. I want to see mankind develop a world that will be at least as good to live in a hundred years hence, as it is today. I am not giving way to despair and I believe in working to change what I see as the present trends; but I cannot be an optimist."[11]

This concern is also shared by the President of the US. In sending the first annual report of his Council on Environmental Quality to Congress, President Nixon stated that "our environmental problems are very serious, indeed urgent, but they do not justify either panic or hysteria. The problems are highly complex, and their resolution will require rational, systematic approaches, hard work and patience." There is no point in attempting to conceal the problems and there must be an open and frank discussion of all the possible consequences and not least the problems of the human internal environment, which may affect us all in some degree, and may have even more serious consequences for the health of future generations, if they are not evaluated with a full sense of urgency.

REFERENCES

[1] *Nutrition against Disease: environmental protection*, Roger J. Williams, Pitman, London, 1971.

[2] *A History of Europe*, H. A. L. Fisher, Edward Arnold, London, 1949, p. 11.

[3] "Agriculture: the roots of deterioration", Walter Ryder, *New Scientist*, 1972, **54** (8th June), 567.

[4] Annual Review of the UN Food and Agricultural Organisation (FAO) reported in the *Financial Times*, 21 November 1972.

[5] "Related Disease-Related Cause", D. P. Burkitt, *Lancet*, 1969, 1229, and *Cancer*, 1971, **18** (1), 3.

[6] "Sugar and Disease", John Yudkin, *Nature*, 1972, **239**, 197.

[7] *Diabetes, Coronary Thrombosis and the Saccharine Disease*, T. L. Cleave, G. D. Campbell, and N. S. Painter, John Wright, Bristol, 1969.

[8] "Internal Pollution: our first priority", Peter Beaconsfield, *New Scientist*, 1971 (18th March), 600.

[9] "Control of Chemical Pollutants", Samuel S. Epstein, *Nature*, 1970, **228**, 816.

[10] "The Role of Plant Foods in Solving the World Food Problem": proteins", F. Wokes, *Plant Foods for Human Nutrition*, 1968, **1** (1), 23.
[11] "Doomsday Syndrome", John T. Edsall, *Nature*, 1971, **234**, 56.

Suggestions for further reading
Toxic Constituents of Plant Foodstuffs, ed. Irvin E. Liener, Academic Press, New York and London, 1960. (A useful survey of natural toxic constituents of plants, with a chapter on adventitious toxic factors in processed foods.)
Plenty and Want: a social history of diet in England from 1815 to the present day, John Burnett, Penguin, Harmondsworth, 1968. (Describes the changing patterns of food habits during the industrial revolution.)

Appendix

THE PERIODIC TABLE OF THE ELEMENTS

GROUP	I	II	3	4	5	6	7	8			1	2	III	IV	V	VI	VII	0
PERIOD																		
1	1 H																	2 He
2	3 Li	4 Be											5 B	6 C	7 N	8 O	9 F	10 Ne
3	11 Na	12 Mg											13 Al	14 Si	15 P	16 S	17 Cl	18 A
4	19 K	20 Ca	21 Sc	22 Ti	23 V	24 Cr	25 Mn	26 Fe	27 Co	28 Ni	29 Cu	30 Zn	31 Ga	32 Ge	33 As	34 Se	35 Br	36 Kr
5	37 Rb	38 Sr	39 Y	40 Zr	41 Nb	42 Mo	43 Tc	44 Ru	45 Rh	46 Pd	47 Ag	48 Cd	49 In	50 Sn	51 Sb	52 Te	53 I	54 Xe
6	55 Cs	56 Ba	57* La	72 Hf	73 Ta	74 W	75 Re	76 Os	77 Ir	78 Pt	79 Au	80 Hg	81 Tl	82 Pb	83 Bi	84 Po	85 At	86 Rn
7	87 Fr	88 Ra	89** Ac															

* LANTHANONS RARE EARTHS	58 Ce	59 Pr	60 Nd	61 Pm	62 Sm	63 Eu	64 Gd	65 Tb	66 Dy	67 Ho	68 Er	69 Tm	70 Yb	71 Lu

** ACTINONS	90 Th	91 Pa	92 U	93 Np	94 Pu	95 Am	96 Cm	97 Bk	98 Cf					

Index

Key:

ADI	Acceptable Daily Intake	Hg	Mercury
As	Arsenic	I131	Iodine-131
Cd	Cadmium	OR	Observed ratio
CF	Concentration factor	Pb	Lead
Cs137	Caesium-137	Sr90	Strontium-90
Cu	Copper	Zn	Zinc

PRINTED IN GREAT BRITAIN BY
COX & WYMAN LTD
LONDON, FAKENHAM, AND READING